D1681706

THE DEVIL YOU KNOW

ALSO BY CHARLES M. BLOW

Fire Shut Up in My Bones

THE DEVIL YOU KNOW

A BLACK POWER MANIFESTO

CHARLES M. BLOW

HARPER

An Imprint of HarperCollinsPublishers

THE DEVIL YOU KNOW. Copyright © 2021 by Charles M. Blow. All rights reserved. Printed in the United States of America. No part of this book may be used or reproduced in any manner whatsoever without written permission except in the case of brief quotations embodied in critical articles and reviews. For information, address HarperCollins Publishers, 195 Broadway, New York, NY 10007.

HarperCollins books may be purchased for educational, business, or sales promotional use. For information, please email the Special Markets Department at SPsales@harpercollins.com.

FIRST EDITION

Designed by Leah Carlson-Stanisic

Library of Congress Cataloging-in-Publication Data has been applied for.

ISBN 978-0-06-291466-8

21 22 23 24 25   LSC   10 9 8 7 6 5 4 3 2 1

IN MEMORY OF MY BIG BROTHER,
FREDERICK EDWARD BLOW:
SO SMART, SO CHARMING, SO COOL.
HE FOUGHT THE GOOD FIGHT.

1962–2020

> FOR A NEGRO, THERE'S NO DIFFERENCE BETWEEN THE NORTH AND SOUTH. THERE'S JUST A DIFFERENCE IN THE WAY THEY CASTRATE YOU.
>
> BUT THE FACT OF THE CASTRATION IS THE AMERICAN FACT.
>
> —James Baldwin, *I Am Not Your Negro*

CONTENTS

INTRODUCTION 1

One
THE PAST AS PROLOGUE 11

Two
THE PROPOSITION 29

Three
THE PUSH 65

Four
THE PULL: THE POWER OF DENSITY 119

Five
THE END OF HOPING AND WAITING 147

Six
THE REUNION 189

Notes 211

INTRODUCTION

I CAN'T BREATHE!

—George Floyd, Eric Garner, Black America,
millions of protesters in the United States and around the world

He called out for his dead mother. He called out for his children.

Knee in his neck, cheek to pavement, the life slowly being pressed out of him by an officer whose demeanor was so unbothered that the sunglasses pushed back to the peak of his head were never disturbed.

George Floyd had been killed, casually, callously, in the street, in the full light of day, with witnesses watching, recording, and objecting.

This single act of morbid street theater, this murder in Minneapolis, struck at something in the social consciousness. There had been other killings of unarmed Black people, but this was different. This was depraved. This was disgusting.

There was no way to reason it away, to make it appear justifiable. There was no ambiguity. It was tragic and it was cruel, and it was the cruelty of it that activated people's anger and disgust. It became a catalyst.

Millions of people, hitherto confined to homes by a deadly pandemic and a halted economy, poured into the street, mostly young, mostly white, to assert that Black lives matter, and to demand police accountability and reform as well as racial justice and equality.

A Pew Research Center report in late June 2020 found that 6 percent of American adults said they attended a protest or rally that focused on issues related to race or racial equality in the month preceding the survey. That's about fifteen million people, an astounding number. The percentage of protesters who were white was nearly three times the percentage who were Black. The percentage of Hispanics was higher than the percentage of Black people as well.

People began to talk about the historic protests in the loftiest of terms, labeling them a racial reckoning, one that was long overdue. Racist monuments came down and supportive placards went up. We painted murals on the streets and took down some statues. Companies committed to changing the Black faces on bottles of syrup and bags of rice. Athletes protested and boycotted and race car drivers held a racial solidarity parade. We held a quasi-social-distanced redux of the March on Washington. There were television specials about injustice and expanded coverage of protests. Books about race rose to the tops of bestseller lists.

States like New York and California passed police reform legislation, and scores of individual departments banned or restricted chokeholds and strangleholds and required officers to intervene when their colleagues used excessive force.

But, national progress, even on the issue of police accountability and reform, remained elusive. The slate of police re-

INTRODUCTION

forms passed by the House quickly became bogged down in the Senate.

And to put it plainly: Most of the action amounted to feel-good gestures that cost nothing and shift no power. They create little justice and provide little equity. Even the House bill, with its de minimis slate of reforms, would basically punish the system's soldiers without altering the system itself. It would make the officers the fall guy for their bad behavior while doing little to condemn or even address the savagery and voraciousness of the system that required their service.

The bill stalled as the protests began to dwindle. People were then forced to consider whether many of the people who marched and carried signs were truly committed to Black lives and Black liberation or whether some, deprived of rites of passage, parties and proms, had simply developed a cabin fever racial consciousness, using the protests as congregational outlets, treating them like a social justice Coachella, a systemic racism Woodstock.

Young people could be outside, together, part of something, feel something. For some, the protests were simply a rebellion against isolation and social distancing. The protests became a proxy for a hall pass.

When Jacob Blake, another unarmed Black man, was shot seven times in the back, in front of his children, again in the street in broad daylight, just one state over from where Floyd was murdered, there was no similar outpouring of outrage. The summer was winding down, schools were reopening, and the fashion had faded.

A poll of people in the state found that in the weeks after Floyd was killed, the approval of Black Lives Matter protests among white Wisconsinites was net +22; in the days leading up to Blake's shooting, it was net −5. Among those who were Black or Hispanic, the net approval held steady at +58.

In some cases, white allies even began to center their own maltreatment while protesting rather than the fundamental issue at hand: the treatment of Black people throughout their lives. How dare the police treat these white liberals poorly, unfairly assault or arrest them? For Black people, state violence and injustice are an intrinsic reality; for white liberals, it was a jarring outrage, an assault on their privilege.

For these protesters, the social justice battle for Black lives was converted into a First Amendment battle for free speech and the right to assemble. That became the glue that bound them to the cause.

But in the binding, as is always the case, the precise, particular grievance of Black America is ever in danger of subsumption. The Black battle is not necessarily joined but hijacked, overwhelmed, by the white liberal grievance.

E. D. Mondainé, president of the Portland, Oregon, branch of the NAACP, wrote an opinion piece about the protests in that city for the *Washington Post* under the headline, "Portland's Protests Were Supposed to Be about Black Lives. Now, They're White Spectacle." In it he questioned:

> *Vandalizing government buildings and hurling projectiles at law enforcement draw attention—but how do these actions stop police from killing black people? What are antifa and other leftist agitators achieving for the cause of black equality? The "Wall of Moms," while perhaps well-intentioned, ends up redirecting attention away from the urgent issue of*

INTRODUCTION

murdered black bodies. This might ease the consciences of white, affluent women who have previously been silent in the face of black oppression, but it's fair to ask: Are they really furthering the cause of justice, or is this another example of white co-optation?

And in the end, however protest is performed, for what motivations, it will eventually wane. Outrage is an expensive emotion. It consumes energy like a blaze. At some point, inevitably, the fuel is exhausted. In the afterglow of it all, in the ash, what have we truly gained from this episode beyond displays of performative activism by organizations and allies, people cosplaying Black allegiance, and legislative tokenism that assuages white guilt and attempts to coax Black people into passivity, into quietly absorbing an endless oppression?

The supposed racial reckoning served only to underscore racism's rigidity. Not much changed for Black people. Power didn't shift. But, it must.

On November 3, 2020, in a historic election with record turnout, nearly half the voters cast their ballots for Donald Trump, an unrepentant racist who ran for reelection under the racially coded "law and order" mantra, encouraged police brutality, defended Confederate monuments, and attacked the Black Lives Matter movement. Joe Biden, a long-standing centrist whose failing candidacy during the primaries, it should be noted, was thrown a lifeline by Black people in the South, claimed victory over Trump as both men received a record number of votes.

But, further down the ballot, little changed. In the weeks following the election, control of the Senate had yet to be determined, while Republicans picked up a few seats in the House. Of the eleven gubernatorial races, only one—Montana—flipped from blue to red. And, it was effectively status quo in state legislatures, as Democrats' attempts to retake the majority fell short. As the National Conference of State Legislatures pointed out in the days following the election: "With just two chamber flips so far, it looks like 2020 will see the least party control changes on Election Day since at least 1944, when only four chambers changed hands." While not all the races had been called at this point, it seems clear that Republicans will control a majority of the redistricting that will take place in 2021.

With so much talk of change, the election was shockingly bereft of it.

Furthermore, if exit polls conducted during a pandemic are to be believed, the way people voted offered a deflating counterpoint to the racial diversity of the summer protests. A majority of white people, both men and women, still voted for the sitting president. In fact, Trump garnered a larger share of the nonwhite vote than any Republican since 1960. And, his share of the LGBT vote doubled from 14 percent in 2016 to 28 percent in 2020.

But there was one glimmer of hope that carried with it a powerful possibility: Georgia flipped from red to blue for the first time since Bill Clinton won the state in 1992. And voters pushed both Senate seats into runoffs for Democratic candidates Jon Ossoff and Raphael Warnock, a Black pastor of Ebenezer Baptist Church in Atlanta, where Martin Luther

INTRODUCTION

King Jr. was once copastor. The coalition of voters that made this possible was led by Black voters, who constituted a majority of those voting for Biden in Georgia, according to estimates from the Associated Press's VoteCast.

The success of the Democratic party's gains in Georgia are in part attributable to a rise in the state's Black population. In the early 1990s, Black people constituted a little over a quarter of the state's population; in 2020, they constitute about a third of it. The Atlanta–Sandy Springs–Roswell metro area saw an increase of 251,000 Black people between 2010 and 2016, the largest gain in a metro area during that time period.

It was also due to a massive voter enfranchisement effort led by former Georgia gubernatorial candidate Stacey Abrams, whose group Fair Fight registered 800,000 new voters in the state in just two years. As Abrams told NPR on the eve of the election: "I will say, of those numbers, what we are excited about is that 45 percent of those new voters are under the age of thirty. Forty-nine percent are people of color." Biden carried the state by only twelve thousand plus votes.

With this election, Georgia became the model for how Black people can potentially experience true power in this country and alter the political landscape.

Georgia became proof of concept—the concept that animates this book.

THE DEVIL YOU KNOW

ONE

THE PAST AS PROLOGUE

● ● ●

IF YOU BLACK YOU WERE BORN IN JAIL, IN THE NORTH AS WELL AS THE SOUTH. STOP TALKING ABOUT THE SOUTH. AS LONG AS YOU SOUTH OF THE CANADIAN BORDER, YOU SOUTH.

—Malcolm X, "The Ballot or the Bullet"

On a cold October day I drove to the South Side of Chicago to meet the man some called the griot. I arrived in Hyde Park, on Drexel Boulevard, a grand parkway designed by Frederick Law Olmsted, the same man who designed New York's Central Park. The street was lined with stately buildings and imposing mansions, some worn, some refurbished, giving the aura of a district aiming earnestly to reset and recover, one trying to reclaim a bygone prosperity that had given way to hope and aspiration, memory and longing, an angst in the air.

There were several apartment buildings on the street. One was the address I sought. I climbed a few flights of cramped stairs and knocked on the door. A woman answered. She appeared to be about the age of my mother, but she had obviously matured with grace and care, a beauty stubbornly resisting the ravages of aging, that touch of timelessness darker skin can bestow. She offered me a seat—I happily took it— and a drink—I respectfully declined.

Her name was Zenobia, "Life of Zeus" in Greek, name of the swarthy third-century Syrian queen. But here, in this life, she was wife of the griot.

The apartment was quiet and brimming with books, laid flat, stacked like towers. On the tables, on the floor, in the corners. But, the clutter didn't reduce the space; it consecrated it.

These were not ornamental objects. These were the prodigious accumulations of a life consumed by an intense desire to know and an unceasing pursuit of enlightenment. This was more shrine than flat, and I shrank in awe of it.

Zenobia disappeared into a back room and reemerged guiding an arm—thin, twisted, brown, like the leafless branches of the trees lining the walkways outside. She was supporting a bent man with horn-rimmed glasses and vanishing hair, blanched by time. His name was Timuel Black—author, educator, activist, and historian—and it was less than two months before his one hundredth birthday. Everyone called him Tim. When Barack Obama came to Chicago, it was Tim who had introduced him to people like Father Pfleger and Jeremiah Wright. As Tim told *Chicago* magazine, "People would trust him because they trusted me."

As Tim approached, I rose, not only out of professional courtesy but also out of ancestral deference. I shook his hand. It was cold and veiny, light and knowing, the way old people's hands are, barely there, half spirit.

He lowered himself deep into a chair, and I began by asking him about what came to be called the Great Migration: the mass migration of millions of African-Americans largely from the rural South to cities primarily in the North and West that spanned from 1916 to 1970. It happened in two major waves, their beginnings roughly corresponding to World Wars I and II, when northern and western factories needed workers to replace men who had gone off to fight.

Before then, 90 percent of all Black people in America lived in the South. While some of the relocation took place within the South itself, the main destinations for these migrants who moved in search of relief from oppression and for the promise of opportunity were cities like Chicago, New York, Philadelphia, Saint Louis, Detroit, Pittsburgh, Indianapolis, Los Angeles, San Francisco, Oakland, Phoenix, Seattle, and Portland—hereafter referred to as "destination cities."

There had been many reasons following emancipation for Black people to flee the South, but they hadn't done so. Even during slavery, a majority of free Black people lived in the South, not the North. They didn't leave. In fact, in the years after the Civil War, the center of Black America moved farther south, not north. The Reconstruction South was revolutionary and the center of Black power at the time. Even as Reconstruction collapsed, Jim Crow rose, and lynchings surged, they had remained.

The only major exodus during that period was the migration of roughly 27,000 black people, mostly from Tennessee, Mississippi, Louisiana, and Texas, who moved to Kansas and other western states during the 1870s, culminating in the so-called Kansas Fever Exodus of 1879. The migrants, who came to be called Exodusters, were drawn by the Homestead Act of 1862 and the possibility of relief from oppression, much like the Great Migration that would follow it, though Exodusters' numbers would pale by comparison. Notably, Frederick Douglass opposed the exodus, contending in part that the South was "the best locality for the Negro on the ground of his political powers and possibilities."

Then, it all changed.

A first question was the only prompt Tim needed. The words flowed out of him easily. His body was frail, but his mind sharp and his voice a booming baritone, gravelly and weighted with gravitas. The contrast was surprising and a bit inspiring. I sat quietly, recording and taking notes, only occasionally interrupting with a question for clarification or redirection and to insert my next inquiry.

The story Tim told me was of one man's journey, of one man's arrival and existence in Chicago, his own, but there was a universalness to its particularness. He spoke for the movement itself, from the very beginnings of it, as well as for himself. He spoke life into history.

He was born in December 1918, son of a "badass nigga." He laughed at his own characterization of his father. By the end of the next summer his family migrated to Chicago to escape the viciousness of the South.

There were quite a few such men making their way north, but Tim's father, married with children, was more the exception than the rule in the early years of the migration. The year Tim and his family migrated, the US Department of Labor's Division of Negro Economics issued a report titled "Negro Migration in 1916–17," which detailed that "the outstanding fact of the Negro migration from the South is that the movement is preponderantly one of single men. Certainly 70 or 80 percent of the migrants are without family ties in the North."

Many of the migrants during the first wave of the Great Migration, in addition to those drawn from the farmlands, came from the urban areas of the South—Birmingham, New Orleans, Charleston—and they fled for multiple reasons: primarily the push of escaping violence and the pull of economic possibility, but also for the promise of better educations for themselves and their children, and to claim the right to vote, a right severely obstructed in the South. This exodus

THE PAST AS PROLOGUE

was accelerated by hundreds of thousands of Black veterans returning from World War I, some having dodged bullets, some having taken them, risking their lives for a country that devalued those lives.

The Black population of Chicago more than doubled during that first migratory wave. Some migrants traveled by bus. Many others traveled by train, the Illinois Central Railroad, which sent trains up from New Orleans and Mobile, through Jackson and Memphis, and right into downtown Chicago's Central Station on Twelfth Street. The trains came to be known derisively as the "chicken bone express" as Black people making the journey often boarded with home-cooked food for the trek but disembarked leaving the car floors littered with bones.

Relatives or friends, often those who'd sent letters encouraging them to come north, would meet the migrants at the station. They would serve as the new arrivals' counsel and chaperones, instructing them on how to survive and behave in the frigid city on the lake. Don't talk loud. Don't spit on the sidewalks (paved walkways were new to many). Dress up for work. If you're reading the *Chicago Defender*, hide it inside one of the white newspapers, the *Chicago Tribune* or the *Chicago Examiner*.

Part of the mission of the newly founded National Urban League was to facilitate this assimilation of southern migrants to the North. During the time of Tim's family's migration, the Chicago Urban League published a six-point "Self-Help" pamphlet for migrants with admonitions like "Do not loaf," "Do not carry on loud conversations in street cars and public places," and "Do not send for your family until you get a job."

There were widespread concerns that newly arrived Negroes not mistake their newly acquired liberty for unbridled license, not only for their own health and safety but also to safeguard the standing and racial equilibrium that existing Blacks in the city believed they had achieved with their white counterparts.

Even some of the most enlightened Black people at that time clung to the specious notion that racial oppression could be managed—aggravated or defused—by the conduct and comportment of the target. The same year that Tim's family migrated, the dean of the College of Arts and Sciences at Howard University, a historically Black college chartered by Congress in the nation's capital two years after the Civil War, wrote a letter to the editors of the *New York Times* in which he warned: "Should the influx of negro laborers to the North, without proper restriction and control, be allowed to prejudice public opinion and thus reproduce Southern proscription in the Northern States, the last state of the race would be worse than the first."

Racial tensions were inflamed when Black people tried to move out of the "Black Belt," a string of neighborhoods on the South Side of Chicago, and into white neighborhoods. The pressure from the swelling ranks of Black migrants and their desires to find decent housing played a part in the bloody riot of 1919 in Chicago that killed thirty-eight people, twenty-three Black and fifteen white. Indeed, that riot was one of many around the country during that period in which violent white mobs attacked Black citizens, who in many cases valiantly fought back. So much blood flowed in America's streets that year that James Weldon Johnson, field secretary of the

THE PAST AS PROLOGUE

NAACP, dubbed it the "Red Summer." In ten months, an estimated 250 people were killed—including nearly 100 who were lynched—in riots in at least twenty-five cities. If you expand the definition of Red Summer to begin with the 1917 riots in Saint Louis and end with the 1923 Rosewood Massacre in Florida, the toll mounts to more than 1,100 deaths.

Three years after the Chicago riot the city's Commission on Race Relations issued a report titled *The Negro in Chicago* that analyzed the riot and its causes. The report concluded that "practically no new building had been done in the city during the war, and it was a physical impossibility for a doubled Negro population to live in the space occupied in 1915." As a direct result, its authors pointed out, "Negroes spread out of what had been known as the 'Black Belt' into neighborhoods near-by which had been exclusively white," and that movement "developed friction, so much so that in the 'invaded' neighborhoods bombs were thrown at the houses of Negroes who had moved in, and of real estate men, white and Negro, who sold or rented properties to newcomers."

White people in Chicago found a way to formalize and ensure segregation: restrictive covenants. These covenants were contractual arrangements between sellers and buyers attached to parcels of land that prohibited Black people from using, occupying, buying, leasing, or receiving property in those areas. The Chicago Real Estate Board not only proposed explicit housing segregation by race, but also petitioned the city council to pass an ordinance prohibiting further migration of Blacks to Chicago until such a time as the city could work out "reasonable restrictions" sufficient to "prevent lawlessness, destruction of values and property, and loss of life."

After the 1919 riots the use of restrictive covenants surged. By 1939, an estimated 80 percent of all Chicago's land area was covered by these covenants.

Many had forfeited land in the South in order to migrate, only to be forbidden from acquiring land where they settled in the North. As James R. Grossman wrote in *Land of Hope*, "To many black southerners, northward migration meant abandoning the dreams of independence through land ownership that had been central to southern black culture since emancipation."

In the abstract, when there were few Blacks in northern cities, people there could look down their noses at the racists in the South. But, when hundreds of thousands of Black people showed up, those northerners had to live up to their ideals. They didn't. Instead, they employed many of the same brutal tactics—oppressive policing, housing discrimination, restrictive employment—that southern racists had used to keep Black folks subordinate and separate. The North's liberalism was marbled with contradiction.

And yet, because of the quality and ability of the migrants and their forced geographic concentration, the citizens of Chicago's Black Belt were in control of its economic and political structures, and in 1929 produced the first African-American elected to Congress since Reconstruction and the first elected outside the South, the Alabama-born Oscar Stanton De Priest.

That segregated concentration also meant that the most successful could be present-in-the-flesh role models for others, a benefit Tim's mother consistently invoked, pointing to the doctor or lawyer and telling young Tim, "You can be just like him."

THE PAST AS PROLOGUE

But in an ironic twist, the turning point—the point at which the many benefits of the Great Migration began to be converted into burden—was sparked by a judicial victory for Black people: the 1948 striking down of racial covenants by the US Supreme Court in the case of *Shelley v. Kraemer*, a case originating from Saint Louis. People with the means to move out of the Black Belt did so almost immediately, bound for whiter communities—too quickly, in Tim's estimation.

These elite Blacks abandoned the sequestered Black communities where they were exalted and essential in an effort to merge into whiteness, where they were tolerated at best, and spurned at worst. It was *A Raisin in the Sun* in real time. The play—the first by a Black woman to be staged on Broadway—was written by the noted playwright Lorraine Hansberry, who knew of which she wrote. Raised in Chicago, her family had been party to the 1940 Supreme Court case *Hansberry v. Lee*, which helped lay the groundwork for overturning racial covenants. Tim had been the Hansberrys' grocery boy.

As Hansberry explained in her posthumously published autobiography, *To Be Young, Gifted and Black*, she and her brothers were taught that "above all, there were two things which were never to be betrayed: the family and the race." And yet these Blacks who fled their own communities for white ones were guilty, inadvertently, of just this betrayal.

"We, who had the skills and the experience" to help provide the economic security that Black people needed "did not share that with our less fortunate brother and sister" as had been done for them, Tim told me. Just their presence in the Black community conferred some advantage by virtue of adjacency.

And this second fracturing of the Black community further

weakened it. The dispersal of people was a direct line to dilution of power, in all its forms, especially political. As Tim succinctly put it, "Segregation enhanced both our knowledge and behavior and opportunity because of the concentration of power—political and economic power.... There was a concentration of political behavior and knowledge that those who were elected had a feeling of responsibility to that group."

And when concentrated Black power in destination cities dimmed and the less-well-off Blacks were left exposed and without powerful champions, the Black communities in those cities descended into strife and sometimes desolation. Bryan Stevenson, the author of *Just Mercy*, the executive director of the Equal Justice Initiative, and the founder of the National Memorial for Peace and Justice, calls migrants of the Great Migration "refugees and exiles of terror." By extension, the Black communities in these cities, abandoned by the most powerful Blacks and spurned by whites, became, functionally, a form of permanent refugee camps.

At the same time, the Great Migration did to the Black South what the transatlantic slave trade had done to West Africa: it drained it of its young and vibrant, stunting its growth and reshaping its culture. Most Africans taken into slavery were young, many even children. Researchers believe that over a third of enslaved transports during the eighteenth century were under fifteen. And, most of those stolen from Africa's shores were men.

This hit the continent like a bomb. At the beginning of the transatlantic slave trade, Europe's population was roughly one and a half times the size of Africa's. By the time it ended,

Europe's population was more than two and a half times the size of Africa's.

Similarly, the Great Migration hit the South like a bomb, siphoning off many of the youngest, brightest, and most ambitious. According to Census data, when the Great Migration began, Black people made up 55 percent of the population of South Carolina. When it ended, Black people composed about 30 percent of that state's population. Six million people would leave the South for the North and West over the course of six decades.

The communities these mostly young, often single Black people created up north, like Chicago's Black Belt, were dynamic and unmoored by the moral rigidity of elders and unencumbered by the need to care for children. That worked to the good and the bad. The cities where they settled became cauldrons of creativity and opportunity and relative prosperity, but the influx strained those cities, put downward pressure on wages, created a scarcity of suitable housing, and packed powder kegs full of the young and restless.

A 2015 study published in the *American Economic Review* found that "migrating North increases age-specific mortality rates by about half for women and somewhat less than half for men," and causes of death for those migrants were all linked to vices like smoking and drinking.

The Great Migration distorted the age curve of Black communities in both the North and the South. The shift to the North of so many young people left the South disproportionately populated by the very old and the very young, with a sunken place in the middle.

The movement also tilted the gender balance in both places:

from the beginning of the second wave of the migration in 1940 to its end in 1970, the birth rate of unmarried Black mothers tripled. There were multiple causes and contributors to single Black motherhood: the rising tide of mass incarceration that followed the Great Migration, disappearing millions of young Black people—most of them marriage-age Black men—from their communities; employment patterns and what the Brookings Institution calls a "reproductive technology shock": the rise in the availability of abortion and contraception that led to dramatic decreases in shotgun weddings and single mother stigma; and the employment and educational opportunities for Black men. But migration trends cannot be discounted.

I was born in 1970 in Louisiana at the end of the Great Migration into a world shaped by vacancy. The landscape was specked by empty houses; there was more work than could be done, more jobs than could be filled. My tiny hometown of Gibsland lost a full quarter of its population between the 1910 and 1920 censuses, during the first wave of the migration. By the time I was born, there were clearly more women than men, and the most recent census estimates there remain three women to every two men in town.

The migration also cleaved the Black community culturally. It created divergent tracks for divergent factions and weakened ancestral bonds and transgenerational continuity, which no amount of family reunions or sending the kids down south in the summers could fully rectify. The sudden loss of so many young Black people meant that for many children, grandparents became primary parents. Southern Black society became in some ways a gerontocracy, its mood

THE PAST AS PROLOGUE

and mores guided by the aged. I didn't go to day care as a child; I went to gray-care. My great-uncle was my babysitter while my parents went to work, and he took me each day to visit his elderly friends. And I wasn't alone. A 2015 study published in *GrandFamilies: The Contemporary Journal of Research, Practice and Policy* observed that during the migration, children who were left in the South "were closely connected to their grandparents and other relatives in the extended family, spending time with them and being exposed to cultural traditions."

Children raised in this type of sage-abundant, exodus-resultant environment are shaped in particular ways, pushed to opposite extremes: either growing more rooted and less restless, more patient and more persistent, or, conversely, stewing over abandonment and loss of parental intimacy. I was one of the former.

There is no doubt that the movement, particularly in the early years, was an economic boon for migrants. They escaped the harshest, most unfair work practices in the rural farming communities for urban areas in the North and West, where wages were higher and pay more fairly meted out.

A 2017 University of Michigan study of Black people who migrated in the 1930s and 1940s found that "compared to a group that did not leave the South, the children of families who left the South graduated from high school at a rate 11 percent higher than their counterparts, made about $1,000 dollars more per year in 2017 dollars and were 11 percent less likely to be in poverty."

Migrants also found some degree of relief from the racial

terror ravaging the South, where they were restricted to menial work with low, if any, pay, and were often obstructed from voting. This is not to mention the omnipresence of lynching: the period between 1877 and 1950 saw over four thousand African-Americans lynched, at least one a week on average, for some ostensible violation of the racial hierarchy.

And yet the initial benefits of the Great Migration have given way, in many ways, to a stinging failure, a dashing of hopes, and an overwhelming feeling of betrayal. Unrest in these former destination cities, more than in southern cities, is continuously exhaled as a roar of disappointment, a sometimes-violent reflex to a broken promise. We saw it manifest in the first wave of Black Lives Matter protests in 2014 and 2015 in places like Ferguson and Baltimore, in Chicago and New York, and in a second wave in the summer of 2020 in cities around the country, including southern cities, and indeed around the world.

But even the tone and tenor of protests, I believe, can diverge by people's connection to land and lineage. Generally speaking, Black people in destination cities have a more transient attachment to place. There, family lineage is often only a few decades old, if that, and many young people are newly transplanted, their sensibilities rooted more in urbanity than in legacy. They are relatively young communities with constant infusions of new energy, volatile energy that pulses.

The South lacks that level of combustible dynamism. The people who remained in the South are more likely to continue to stay put. Take my home state of Louisiana, for example—nearly four out of five residents were born in the state, a figure that hasn't changed much over the past century.

THE PAST AS PROLOGUE

In Louisiana, and in the South writ large, race must be reckoned with head-on. In large measure, your townspeople, white or Black, were your mother's townspeople and will be your children's townspeople. This forces some form of solution, or at least a détente; the alternative is never-ending enmity. Racism doesn't wither, but is trained when to advance or retreat. It becomes self-regulating. Black and white people in the South have ever been simultaneously at war and in relationship, a fierce conflict winding toward fragile covenant.

But no matter how people, regardless of region, contend with racism—through protest and rebellion, or backyard-fence rapprochement—the racism itself remains.

The conclusion I have come to is simply this: racism behaves the way racism behaves. Racism wasn't and isn't geography dependent, but proximity-and-scale dependent. Black people fled the horrors of the racist South for so-called liberal cities of the North and West, trading the devil they knew for the devil they didn't, only to come to the painful realization that the devil is the devil.

As Julian Bond once put it, "America, after all, unscrambled, spells 'I am race.'"

I often think of racism as having developmental cycles. In the South, it's an old man. There, racism hasn't vanished (far from it), but it has come to terms with itself. In the North, particularly in destination cities, racism is a teenage boy, acting out as the old man did years ago.

TWO

THE PROPOSITION

❃ ❃ ❃

WHAT WE ADVOCATE IS THE MIGRATION OF
LARGE NUMBERS OF PEOPLE TO A SINGLE STATE
FOR THE EXPRESS PURPOSE OF EFFECTING
THE PEACEFUL POLITICAL TAKE-OVER OF THAT
STATE THROUGH THE ELECTIVE PROCESS.

—James F. Blumstein and James Phelan,
"Jamestown Seventy"

The proposition is simple: as many Black descendants of the Great Migration as possible should return to the South from which their ancestors fled. They should do so with moral and political intentionality. And as many Black people not descendant of American slavery as possible should join in their resettlement.

Consider this: In the first census after the Civil War, three Southern states (South Carolina, Mississippi, and Louisiana) were majority Black. In Florida, Blacks were less than 2 percentage points away from constituting a majority; in Alabama, it was less than 3 points; in Georgia, just under 4.

But, as previously noted, during and after the Great Migration the percentages plummeted. Black births in those states barely kept pace with the departures, so the Black population was stagnant as the white populations boomed. In Alabama, for instance, the population was roughly evenly split between Black and white in 1880; when the migration ended in 1970, whites in the state outnumbered Blacks three to one. This was true for the South as a whole. From 1910 to 1970, the Black population in the South grew by only 36 percent; the white population swelled to two and a half times its 1910 size.

Now imagine an alternate scenario: if the Great Migration hadn't happened, and those Black people had remained in the South until the passage of the Civil Rights Act of 1964 and the Voting Rights Act of 1965, it is possible that African-Americans would dominate the politics of the Deep South. They could control or be the driving force in electing as many as twelve US senators.

Considered another way, 44 percent of Black people in America now live outside the South. However, hypothetically speaking, if just half of them moved back south and were strategically arrayed, it would be enough to make Black people the largest racial group in Louisiana, Mississippi, Alabama, Georgia, and South Carolina, a contiguous band of Black power that would upend America's political calculus and exponentially increase Black political influence.

There were only three Black US senators in the 116th Congress, only one of whom is from the South: Republican Tim Scott of South Carolina, who received a scant 8 percent of the Black vote in that state when he ran in 2016. The other two were Kamala Harris of California, where Black voters formed only 6 percent of those voting in 2016, and Cory Booker of New Jersey, where Black people constitute 13.1 percent of those voting, according to exit polls.

All three of those Black senators were elected by white people, meaning white people were either a majority or plurality of the coalitions that swept them into power, according to exit polls. In fact, as of 2020, in the entire 230-year history of the Senate, Black people have never popularly elected a senator on their own or been the majority of a coalition that did so. There have been only ten African-American senators, and they have all been either appointed or elected by white people.

THE PROPOSITION

This is not problematic in theory, but as is the case with all politics, politicians are most beholden to the electorate that installs them.

Furthermore, if the Great Migration hadn't taken place, Black people could control or form the majority influence for as many as ninety Electoral College votes, more than California and New York State combined. And, if they and other groups voted the same way that they now do, they could have ensured that almost every president in the last fifty years was a Democrat.

More specifically, if in the 2016 presidential election Hillary Clinton had been able to carry the states in which Black people are already the majority or plurality of voters in the Democratic primaries—Louisiana, Mississippi, Alabama, Georgia, and South Carolina—she would have become president, midwestern losses notwithstanding, and would have added three new justices to the Supreme Court.

View this plan for Black power in the context of the current landscape of white power. White people have constituted a majority of the population in every state but Hawaii for the last ninety years. Eight states—Maine, Vermont, New Hampshire, West Virginia, Idaho, Wyoming, and Iowa—are over 90 percent white and control one out of every six senate seats in America. The Black population is four times the population of those eight states combined but controls no senate seats.

Not one of these states under white people's control has erased white supremacy. Is it outlandish for Black people to seek to do what their white countrymen have failed or refused to do, and lift state oppression?

There is a precedent for this idea. It grew out of the hippie movement of the 1960s and '70s and its opposition to the Vietnam War.

"If a vocal minority, however fervent its cause, prevails over reason and the will of the majority, this nation has no future as a free society," Richard Nixon proclaimed during an Oval Office address on November 3, 1969, nearly a year to the day from when he was elected.

The warning was targeted at the raging protests against the Vietnam War. A little over two weeks earlier, on October 15, Vietnam moratorium protests had engaged an estimated two million people in cities and towns across the country. As a candidate Nixon had promised that he had a "secret plan" to end the war. When elected, however, he changed his position. Thousands of soldiers were still dying.

"In January I could only conclude that the precipitate withdrawal of American forces from Vietnam would be a disaster not only for South Vietnam but for the United States and for the cause of peace," he read from a prepared text, glancing down frequently at his desk. He asked for the support of the "silent majority" for his position. Indeed, the address would become known as the "silent majority" speech.

Nixon went on to announce his support for the "Vietnamization" of the war, a plan that involved training South Vietnamese soldiers to eventually replace American ones. The plan had been underway for months. As Nixon put it: "In the previous administration, we Americanized the war in Vietnam. In this administration, we are Vietnamizing the search for peace."

This speech did nothing to squelch the antiwar fervor. To the contrary, less than two weeks later the Vietnam Moratorium Committee staged what is believed to be the largest single antiwar demonstration in American history, with more than 500,000 people converging on the Mall in Washington, DC, demanding a rapid withdrawal from the war. As the *New York Times* described the gathering: "Overall, it was a mass gathering of the moderate and radical Left, including the 100 organizations that make up the New Mobilization Committee to End the War in Vietnam, sponsor of the demonstrations, old-style liberals; Communists and pacifists and a sprinkling of the violent New Left."

But while protests are a powerful form of statement-making, and narrative-shifting, their record is less clear on policy advancement. Putting one's boots in the street is an immediate action, unfettered by the creep of legislation or the periodicity of elections. But, as the antiwar movement was forced to realize again and again (and as we've been reminded in recent years), as invigorating and unifying as they are, protests don't stop the mechanisms of power from proceeding as planned. Protesting is a form of direct democracy, but in America national policy is made by representative democracy.

On November 26, just eleven days after the protest, the Draft Lottery Bill was signed, and less than a month after the "silent majority" speech, the United States held its first draft lottery on December 1.

As the war intensified, so did the movement against it. By the early part of 1970, the Weather Underground, an offshoot of Students for a Democratic Society, had claimed credit for "25 bombings—including the U.S. Capitol, the Pentagon, the California Attorney General's office, and a New York City police station," according to the Federal Bureau of Investigation. The group was protesting not only the war but also racism.

On April 30, 1970, Nixon stood before a map of Cambodia during another televised address to announce that American troops would invade Cambodia to cut off supply chains into North Vietnam.

It was two years before a somewhat radical plan was floated in an unlikely venue; while it wasn't directly related to ending the war, it addressed the need to acquire power to craft legislation that would allow citizens to oppose it. The April 1972 issue of *Playboy* featured an article by the writer Richard Pollak entitled "Taking Over Vermont," in which Pollak proposed:

> *Suppose the nation's alienated young decided to stage a takeover of Vermont. Not by staging a weekend rock festival at Rutland and then hanging around the Green Mountains like freaked-out trolls. Not by lacing the water supply with assorted chemical brain scramblers. Not even by trashing the 14-kt.-gold-leaf dome off the Statehouse in Montpelier. Suppose they decided to do it within the system, the hard-hat-approved American way—by ballot!*

Pollak then proceeded to lay out the arithmetic:

The 1970 census counts 444,732 bona-fide residents of the state of Vermont. Of that number, only 287,575 are 18 years old or over and thus eligible by state law to vote in state and local elections. Since 107,527 of these eligibles are between the ages of 18 and 34 and, figuring conservatively, one third of them (35,806) would be likely to sit down and break grass with all incoming pilgrims, the potential enemy strength reduces to 250,000. Lop off another ten or so percent for those good citizens who wouldn't bother to exercise their franchise, even at the prospect of a Yippie governor, and the numerical tipping point comes down to 225,000, give or take a Yankee.

Pollak's article sprang from an obscure 1971 paper entitled "Jamestown Seventy," published in the *Yale Review of Law and Social Action* and penned by two law students, James F. Blumstein and James Phelan. In it, the authors set forth a tantalizing idea: While they believed revolution was the answer to what ailed America, they conceded that it was "impossible when armed revolt by the citizenry-at-large would inevitably be put down by the military might at the disposal of those in control."

But then the authors put forth another vision of revolution:

We see the best way out in rededicating this nation to its heritage: reopening the frontier, where alienated or "deviant" members of society can go to live by their new ideas; providing a living laboratory for social experiment through Radical Federalism; and restoring effective political communication in a multimedia society.

What we advocate is the migration of large numbers of

people to a single state for the express purpose of effecting the peaceful political take-over of that state through the elective process.

In fact, the ground for a takeover had been seeded well before *Playboy* published Pollak's article; between the years 1965 and 1975, 100,000 young, like-minded people would move to the state. In the decades that followed, Vermont was transformed by "a generation of hippies and free thinkers . . . using the tried and true American way, by the ballot, electing more progressive politicians per capita in the intervening decades than any state and enacting legislation that routinely resulted in Vermont ranking among the most liberal states in the union," in the words of the writer Yvonne Daley.

Patrick Leahy, now the longest-serving senator in America, was elected as Vermont's first Democratic senator in 1974. The influx no doubt also propelled the progressive independent Bernie Sanders, who had moved to the state in the 1960s, to become mayor of Burlington in 1981, before moving on to the US House in 1991 and then the Senate in 2007, where he caucuses with Democrats.

It should be noted that Vermont is one of the whitest states in the union, with just 5.8 percent of the population identifying as nonwhite, but in 1970 the state was even less diverse, with the nonwhite population measuring a meager 0.4 percent. I'm proposing that, in the same way that young white liberals moved en masse to one of the whitest states in the country with a political intent, young Black people move to the Blackest states in the union to amass political power.

THE PROPOSITION

The Great Migration had some detractors in the Black community, but it also had many champions, including political leaders, intellectuals, and writers. Black journalists played a significant role in encouraging and promoting the movement.

Perhaps no one person's influence surpassed that of the Georgia-born Robert Sengstacke Abbott, publisher of the *Chicago Defender*, the leading Black newspaper of the time. Abbott was a major proponent of the migration, which he thought would not only better the lives of Black people but also punish the white population of the South for their poor treatment of Blacks.

Abbott came to call it the "Second Emancipation." His *Defender* was indispensable in encouraging, chronicling, and providing a moral and civic frame for the Great Migration. The newspaper was often taken south, sometimes smuggled like contraband, by the Pullman porters on the trains, and it served as a welcoming beacon for those considering the trek. Abbott's *Defender* was heroically heretical, like Martin Luther's theses nailed to the door of the white South.

It is my great hope and ambition to stand on Abbott's shoulders, to do for the reverse migration what he did for the Great Migration.

I am not an activist, and I have never thought of myself as a revolutionary. I am a newspaperman. I bear witness. I interpret the world. I record history in real time.

In fact, the knock on me from the most progressive, even

radical quarters has been that I'm too much of an institutionalist and corporatist. I guess that to some degree, that criticism is fair. Not only did I intern at the *New York Times*, I've worked there for the past twenty-six years, half of my life, with only a short break to work at *National Geographic*. That wasn't my plan. I didn't have a plan. It is just the way life worked out.

Newspapering is in my blood. When I was a boy, no matter how poor we were, one of the things my mother would never forgo was a home subscription to the local newspaper, the *Shreveport Times*. She read it front to back, and she saved the occasionally printed "Mini Page" for me. In high school, I started a school newspaper and began to write letters to the editor of the *Shreveport Times*, a couple of which they printed. In college, I became editor of the school newspaper, *The Gramblinite*, and started an accompanying magazine, *Razz*, which I wrote, edited, and designed myself. And, before coming to the *New York Times*, I worked at the *Detroit News* for nearly two years. This is my life.

The moment that I realized that I could be a true insurgent in this space came in 2013. I was at the Ford Foundation for an all-day series of lectures called The Road Ahead for Civil Rights: Courting Change. Harry Belafonte addressed the room during the lunch session. He spoke in a low-but-sure raspy voice, diminished by age, but deepened in solemnity. He was elegant, erudite, and searing. I craned my neck to see if he was reading from a prepared text. He wasn't. I was mesmerized; everyone and everything else in the room receded. I don't recall the people with whom I was sitting or anyone's reaction. I was transfixed as if he were speaking to only me.

THE PROPOSITION

And then, having offered the provocative assessment that then-president Barack Obama had "suffocated radical thinking," he posed a question: "Where are the radical thinkers?"

Belafonte is the epitome of the artist-activist, having risked not only his career in service of social justice, but also his life. In the summer of 1964 he received an urgent call from Jim Forman, the executive secretary of the Student Nonviolent Coordinating Committee, explaining that the young people involved in Freedom Summer in the South, an effort for which Belafonte had already raised money, needed an additional $50,000 to extend their campaign into the fall.

He raised $70,000, but knew he'd have to deliver it himself. He asked his best friend, Sidney Poitier, to travel with him to make the delivery. They stuffed the money, small bills, into a doctor's bag, and caught a flight to Mississippi. When they landed on the final leg of their journey in Greenwood, there was a row of trucks at the end of the airfield. It was the Klan. They chased Belafonte, Poitier, and their escort down the country road, ramming the back of their car and firing at them from close range. At the time, Belafonte and Poitier were two of the biggest Black stars in America, and even from that perch they risked it all.

On the walk back to the *Times*' Midtown offices, Belafonte's query, "Where are the radical thinkers?" kept replaying in my head, and it occurred to me that I had been thinking too small, all my life, about my approach to being in the world and conceiving my role in it. I had to remember that a big idea could change the course of history. And, I was uniquely positioned, as a writer, not only to express such an idea but also to push it out into the world.

This book is my big idea. I posit that destination cities for many of the participants of the Great Migration and their descendants have become locked in a form of perpetual oppression—geographically, economically, and politically isolated and ignored.

Black people in America should reverse the Great Migration, and return to the states where they had been at or near the majority after the Civil War, and to the states where they currently constitute large percentages of the population. In effect, Black people could colonize and control the states they would have controlled if they had not fled them.

I realize that I am proposing nothing short of the most audacious power play by Black America in the history of the country. This book is a grand exhortation to generations of a people, my people, offering a road map to true and lasting political power in the United States.

The question before Black America is profound: What could and should Black people do to acquire and maintain the economic and political power—for the many, not just for the few—that the Great Migration failed to secure?

Of course questions—and doubts—abound about such a proposal, questions I have anticipated and attempt here to address.

What do you mean by "the South"? Which states?

In many ways, the American South is whatever people believe it is. People generally agree that the Deep South states of Louisiana, Mississippi, Alabama, Georgia, and South Carolina make up the South, but there remains some consterna-

THE PROPOSITION

tion about how far beyond them into the peripheral states to draw the borders. Is the Mason-Dixon Line one of modern demarcation? Should the South be confined to the eleven Confederate states during the Civil War? Is the South purely geographic, or is it also cultural? For instance, while West Texas is most assuredly culturally western, East Texas is culturally southern. (In 2015 John Nova Lomax of *Texas Monthly* magazine wrote that East Texas, where most Texans live, "was every much a part of the King Cotton economy as Alabama or Mississippi. . . . Counties were named in honor of Jefferson Davis, Robert E. Lee, John C. Calhoun, Stonewall Jackson, and John Bell Hood. Most towns of any size sported a prominent monument to its Confederate dead, and up until the early twentieth century, Dallasites and Houstonians saw themselves as just as southern as Memphians or New Orleanians.")

By the same measure, while North Florida is solidly southern, South Florida is culturally not southern at all. As Kyle Munzenrieder of the *Miami New Times* puts it, South Florida is "some mix of New York and the Caribbean."

Definitions of the South can be endlessly debated, but I will defer to the United States Census Bureau's designation, which defines the South as stretching from Texas up to Delaware, and between those two states includes Oklahoma, Arkansas, Louisiana, Mississippi, Alabama, Tennessee, Kentucky, Georgia, Florida, South Carolina, North Carolina, Virginia, West Virginia, and Maryland.

This is admittedly an expansive interpretation, which far exceeds most southerners' sense of the South. (In 2014 the website FiveThirtyEight published the results of a poll in which respondents were asked which states composed the South. Only 6 percent of southerners viewed Maryland as a southern state, and "in addition to Maryland, Oklahoma and West Virginia both pulled less than 25 percent support.")

In point of fact my reverse migration argument targets only nine of the states that are included in the Census's count: Louisiana, Mississippi, Alabama, Georgia, South Carolina, North Carolina, Virginia, Maryland, and Delaware. Specifically, I suggest gravitating to the major cities, arranged like jewels on a chain, which dot the Interstate 20–Interstate 95 corridor from Shreveport, Louisiana, to Wilmington, Delaware. They include Jackson, Mississippi; Birmingham, Alabama; Atlanta, Georgia; Columbia, South Carolina; Charlotte, North Carolina (about ninety miles north of I-20); Richmond, Virginia; Washington, DC (preferably in the Maryland suburbs because the district, unfortunately, is not a state and thus fields no senators); and Baltimore, Maryland.

All these cities, with the exception of Charlotte and Columbia, are majority Black or have a Black plurality. All but one—Wilmington—have a Black mayor. And, four—Jackson, Atlanta, Columbia, and Richmond—are state capitals.

These cities each offer a level of urbanity, for which reverse migrants may long, and also a level of cultural and artistic freedom. Indeed, these cities would do well to view themselves as an interstate confederacy of in-common populations, facing in-common issues, and sharing in-common cause.

The South has long existed as an archipelago for Black people, islands of safety and prosperity interrupted by stretches of hostility and despair. These cities would do well to build bridges of cooperation among them.

I also chose these cities because, inland and distanced from the problematic flood plains, coastal and gulf areas most vulnerable to hurricanes, and tornado alleys of the East and Midwest—not to mention the droughts and wildfires of the West—they would be less affected by the coming climate change catastrophe.

Isn't the proposal racist on its face?

For me, it is by no means racist, because I don't believe race exists in the way many people think it does. I am unapologetically pro-Black, not because I believe in Black supremacy, which is as false and reckless a notion as white supremacy, but rather because I insist upon Black equity and equality. I consider myself an antiracist, and as such I am a fierce warrior against anti-Blackness. In a society and system in which white supremacy is ubiquitous and inveterate, Black people need fierce advocates to help restore the balance, or more precisely, to establish that balance in the first instance.

I concur with the American Anthropological Association's Statement on Race:

> *How people have been accepted and treated within the context of a given society or culture has a direct impact on how they perform in that society. The "racial" worldview was invented to assign some groups to perpetual low status, while others were permitted access to privilege, power, and wealth. The tragedy in the United States has been that the policies and practices stemming from this worldview succeeded all too well in constructing unequal populations among Europeans,*

Native Americans, and peoples of African descent. Given what we know about the capacity of normal humans to achieve and function within any culture, we conclude that present-day inequalities between so-called "racial" groups are not consequences of their biological inheritance but products of historical and contemporary social, economic, educational, and political circumstances.

Race, as we have come to understand it, is a fiction; but, racism, as we have come to live it, is a fact. The point here is not to impose a new racial hierarchy, but to remove an existing one. After centuries of waiting for white majorities to overturn white supremacy, it seems to me that it has fallen to Black people to do it themselves.

Furthermore, my call for Black power through Black majorities isn't intended to exclude white people but rather to expunge white supremacy. Black majority doesn't mean Black only. Even in the three states that held Black majorities that I referenced earlier—South Carolina, Mississippi, and Louisiana—those majorities were far from overwhelming, peaking at 61 percent, 59 percent, and 52 percent, respectively.

And, a majority-Black population doesn't necessarily mean a Blacks-only power structure. There are cities in the Northeast and Midwest that have a Black majority or plurality, like Detroit, Philadelphia, and Saint Louis, and have white mayors. The point is not to create blind racial devotion, but rather to create responsive, race-conscious accountability.

While I espouse no desire for racial exclusion, for white America racial exclusion has been scripture.

Isn't the North just better for Black people than the South?

THE PROPOSITION

Regularly, after the Black people I have spoken to have exhausted all their questions about my big idea, if they've retained a reticence, it is this: they are leery of the South, and often even confess a fear of it.

They still have in their minds a retrograde South: dirty and dusty, overgrown and underdeveloped, a third world region in a first world country. They see a region that is unenlightened and repressive, overrun by religious zealots and open racists. The caricatures have calcified: hillbillies and banjos, Geechees and Bamas, Confederate flags and Ku Klux Klan, backward thinking and religious extremism.

To be sure, all of that is there, and history would justify apprehension. I am the first to admit that there are some cultural hurdles in the South, but I see cultural hurdles, of varying sorts, everywhere. As for racism, I believe it's more evenly distributed across the country than we are willing to admit.

It is true that in surveys, people in the North express support for fewer racially biased ideas than those in the South, but such surveys reveal only which biases people consciously confess to, not the ones they subconsciously possess. So I asked the researchers at Project Implicit, an international collaboration of researchers who use an online implicit association test to measure the racial prejudice of its participants, to run an analysis of their massive data set to see if there were regional differences in pro-white or anti-Black bias. The result, which one of the researchers described as "slightly surprising," was that there was almost no difference in the level of this bias between white people who live in the South and those who live in the Northeast or Midwest. (The bias of white people in the West was slightly lower.)

White people outside the South say the right words but many possess the same bigotry. Racism is everywhere. And if that's the case, wouldn't you rather have some real political power to address that racism? And a yard!

History has a way of drifting toward the most amenable version of itself, one devoid of the disagreeable, and northern racial hypocrites have developed a narrative in which they hold aloft their own halos, insisting that they are the only saints and those in the South the only sinners.

As long as the northern liberals could maintain the illusion of their moral superiority, they could also justify their own lack of progress in terms of racial equality. The North's arrogant insistence that it had no race problem, or at least a minimal one, allowed a racialized police militarism to take root and flourish there. It was a kind of once-removed racism in which individual citizens could keep their hands clean, claiming deniability for the oppression that they passively facilitated. It allowed a balkanized housing and education segregation to develop in supposedly "diverse" cities. It allowed for the rise of Black ghettos and concentrated poverty as well as white flight and urban disinvestment.

The supposed egalitarianism of northern cities is more veneer than core doctrine. It is a flimsy disguise for a racism and white supremacy that diverges from its southern counterpart only in style, not substance.

And, while the North has been stuck in its self-righteous stasis, the savagery of the South has in some ways softened, or morphed. I am careful here not to position progress in the South as fully redemptive or restorative. There are still tremendously oppressive systems operational in the region, cor-

THE PROPOSITION

rupting everything from criminal justice to electoral access. Rather, I simply make the point that there are some positive changes in the region that are undeniable, like the emergence of political power and a thriving Black middle class in many cities. The "New South" is more of an aspiration to an eventual reality than a depiction of an achieved redemption.

But the wishful idealizing of a New South is no more naive than a willful blindness to the transgressions of the Now North. Racism is everywhere. As the celebrated author Jesmyn Ward wrote in 2018 in *Time* about her decision to leave Stanford and move back to Mississippi, American racism is an "infinite room": "The racist, misogynistic sentiment I encounter every day in Mississippi is the same belief that put in place the economic and social caste systems that allowed America to become America. It is the bedrock beneath the soil. Racial violence and subjugation happen on the streets of St. Louis, on the sidewalks of New York City and in the BART stations of Oakland."

Black people have traversed this country in search of a place where the hand of oppression was lightest and the spirit of prosperity was greatest, but have had to learn a bitter lesson: the whole country can be hostile.

Won't this idea encounter powerful opposition, even from liberals?

The fear argument falls flat for me. I am forced to question the people who hide behind it: When has revolution ever been easy? When has a ruling class humbly handed over power or an insurgent class comfortably acquired it? Revolution, even a peaceful one, is frightening, and dangerous, because those with power will view any attempt at divestiture as an act of war.

This furious opposition will likely take many manifestations. There will likely be opposition from the Black elite class in destination cities, people whose power is based on their leadership and influence over a sizable Black population. There will also be opposition from the Black political class whose offices will be in jeopardy if the Black population in their cities shrinks. This is a very real concern, which I want to address forthrightly: There may be some fluctuation in Black political representation during the course of a reverse migration, and, in the beginning, positions added in the South may not balance out those lost in destination cities and states. This is a function of how political machines operate, the way regions are gerrymandered, the way parties horse-trade, the way the establishment grooms ascendant stars, and the way voter suppression is inflicted. But, in the end, the benefit and abundance of Black political power would be to the good.

Even some white liberals, those who call themselves allies, may shrink from the notion of Black power, drawing a false equivalence to the concept of racial superiority espoused by the white power movement. They recoil from the very mention of Black power even as they live out their lives in a world designed by and for white power, not only the hooded and hailing, but also the robed and badged.

Others may simply mourn the notion of a path to Black equality that doesn't feature a starring role for white liberal guilt, that doesn't center on their actions and their capacity for growth and evolution, but skips over it altogether.

Still others may simply hesitate, because on first blush it sounds like throwing in the towel on the grand experiment of multiculturalism. I sought for months to put this proposal

THE PROPOSITION

to Bill Clinton, someone who I thought had deftly navigated the racial minefields in the South. I got my chance in the wee hours of a summer night on Martha's Vineyard. He responded with curiosity and inquisition but not agreement or endorsement. The lack of approval was not deflating, because it had not been requested. Black people need no permission to seek their own liberation. I simply wanted to plant the seed of the plan's plausibility and possibility and watch him wrestle with it.

Does the proposal advocate segregation?

I'm not a strict segregationist any more than I'm a strict integrationist. The former is a self-inflicted injury in a society in which people must operate on some level under the umbrella of multicultural pluralism, but the latter can produce its own wounds when true and full inclusion and equality are denied to minority groups who move out of their homogenous enclaves only to be met by hostilities.

I believe strongly in the value of diversity of all kinds. Diversity makes societies not only more tolerant but more vibrant; social scientists have provided reams of evidence to this effect. In our daily lives, being exposed to people of different ethnicities can reduce the feeling of racial anxiety. In the workplace, diversity can drive innovation, while its absence can hamper productivity.

But, one of the problems with diversity as a living experiment is that white people in America view diversity as a mirroring concept, in which an ideal diversity reflects the demography of the country as a whole: they are the majority, and minorities are present in ratios correspondent to their national share.

In 2018 *Vox* published a report entitled "White America Is Quietly Self-Segregating," in which it quoted Maria Krysan, a sociologist, who concluded from research she conducted in Cook County, Illinois, that people want a specific kind of diverse neighborhood. While the most desirable mix for Blacks and Latinos had them each at about a third of the population, white people's desired mix had them at slightly more than half the population, with all other groups splitting the other half. In white people's vision of diversity, they must still dominate.

It has been shown that living in intimate contact with people who are different from you reduces your racial hostility, but in many destination cities, a strict cordoning off of most people of color means that a white person's only personal and sustained contact with Black people is with the "exceptional Negroes," which creates a kind of false intimacy and the illusion of open-mindedness. Population patterns in these cities are such that liberal whites can experience diversity when and to the degree that they desire (or not), and then retreat safely to their enclaves of homogeneity.

Diversity of the destination cities is a fiction, a deception that prevents the exposure of the lie of the "liberal North." In 2017, *24/7 Wall St.* conducted an analysis of the sixteen most segregated cites in America. Ten of them were destination cities in the North: Detroit, Chicago, Cleveland, Buffalo, Baltimore, Saint Louis, Milwaukee, Philadelphia, Dayton, and Washington, DC. And this had consequences beyond the social:

> In these highly segregated cities, black residents are even more likely to live in extreme concentrated poverty. In seven of the

THE PROPOSITION

cities on this list, more than 20 percent of black residents live in neighborhoods where at least 40 percent of the population is poor. In Detroit, the most segregated major metropolitan area in the country, 1 in 3 black residents live in highly impoverished neighborhoods.

Furthermore, neighborhood segregation has a direct impact on school segregation. The most segregated school system in America is no longer in the South; it's the public school system in New York, its statistics heavily skewed by New York City's school system, the largest and one of the most segregated in the country.

The year I was born, Mississippi senator John Stennis, a segregationist, offered an amendment to school integration policies aimed to call the bluff of northern senators: racial integration policies would have to be uniform across all states. If the white children in the South would have to go to school with Black ones, so would the white children in the North. The amendment was a clever stratagem, meant to expose northern duplicity: that they would reject for themselves that which they prescribed for the South.

Abraham Ribicoff, a liberal senator from Connecticut, took the opportunity to denounce northern racism, declaring on the Senate floor: "The North is guilty of monumental hypocrisy in its treatment of the black man." He went on to say that "Northern communities have been as systematic and consistent as Southern communities in denying to the black man and his children the opportunity that exists for white people." Senator Claiborne Pell, a Rhode Island Democrat, agreed, adding that many northerners "have hypocrisy in our hearts—we go home and talk liberalism to each other, but we don't practice it."

That is as true now as it was then.

We now live in a hybrid America, especially in the North: one of labor integration, but social segregation. We work together. We share mass transportation and public accommodations, and bars, and restaurants. But in our more intimate spaces—where we live and learn—we are resegregating. More precisely, white people are resegregating, self-segregating.

Might a majority-Black state or region experience the same white flight that crippled many cities when they became majority Black?

Possibly, but white flight is wholly different on a state level. It would have the inverse effect of actually strengthening Black political power rather than weakening it. Every state, regardless of composition or population, is constitutionally guaranteed two senators, and has dedicated electors in the Electoral College. The more a state's population shrinks, the more the voting power of those who remain grows.

On average there are about 436,000 people per Electoral College vote. But in sparsely populated states like Wyoming there are 143,000 per electoral vote, while in the heavily populated New York there are 500,000 people per electoral vote. In other words, one voter in Wyoming has the power of nearly four voters in New York. If white people fled states to which Black people returned, that would only strengthen the Black vote.

Furthermore, states have some rigid beneficial realities. They have natural resources and indigenous industries. Many southern states are already heavily subsidized by the federal government. In the 2017 fiscal year, Louisiana and Mississippi trailed only Montana and Wyoming in receiving the largest percent of state revenue from Washington. Louisiana received 43.7 percent of its revenue from the federal government; Mississippi, 43.3 percent. Black people would be in position to regulate the resources and allocate the funds.

Density and state control could also shift the dynamics of our two-party political system and adjust how those parties relate to Black people.

This proposal is not an endorsement of the Democratic Party or progressivism or liberalism, per se. Although I consider myself staunchly liberal, particularly on social issues, most Black people are not as liberal on those issues. It is true that a 2019 Gallup survey found that "aside from Democrats, only two groups show strong liberal tendencies: adults with postgraduate education (15 points more liberal than conservative) and Blacks (nine points more liberal)," and most Blacks identify as Democrats. But, on many social issues—like same-sex marriage—the views of Black people are closer to those of Republicans.

While my own beliefs are liberal, my southern migration proposition is an advocation of Black self-determination in America, whatever form it should take.

Black people, in general, haven't been Democrats by nature, but by necessity. They haven't been as socially progressive as the rest of the party, but they absolutely abhor the Republican abidance of racism. Even among millennials who, regardless of race, are more liberal than their elders, Black millennials are significantly more religious than non-Black millennials, and religiosity informs opinions on social issues.

As NPR's Karen Grigsby Bates reported in 2014, "If you'd walked into a gathering of older black folks 100 years ago, you'd have found that most of them would have been Republican" because it was the "party of Lincoln. Party of the Emancipation. Party that pushed not only black votes but black politicians during that post-bellum period known as Reconstruction."

The Democratic Party was the party of slavery. It was the party of the KKK and the Black codes. It was the party that fought Reconstruction and established Jim Crow. It was the party of white supremacy. It was the party of Alabama governor George Wallace, who declared in 1963, "Segregation now, segregation tomorrow, segregation forever." It was the party of former Klan leader Robert Byrd, who filibustered the Civil Rights Act for fourteen hours.

But allegiances flipped.

The first wave of defections by African-Americans from the Republican to the Democratic Party came with Franklin D. Roosevelt and the New Deal in the 1930s. As Mary McLeod Bethune once explained of the Roosevelt era: "This is the first time in the history of our race that the Negroes of America have felt free to reduce to writing their problems and plans for meeting them with expectancy of sympathetic understanding and interpretation."

By the mid-1930s, most Blacks were voting Democratic, although a sizable percentage remained Republican. Then came the signing of the 1964 Civil Rights Act by the Democratic president Lyndon B. Johnson.

In the face of opposition from southern Democrats, the bill passed in the Democratic-led Congress only with the aid of significant Republican support. In response, Barry Goldwater waged a Republican presidential campaign built in part on his opposition. Goldwater emphatically favored states' rights. As such, he believed the Civil Rights Act unconstitutional. But, since it had become law, he advocated the notion that the pace of implementation should be determined by the states. Whites were reassured by the message; Blacks were shaken by it. (The campaign proved disastrous for Goldwater, but the seeds for future movements were planted; as Bates wrote, "Goldwater can be seen as the godfather [or maybe the midwife] of the current Tea Party.")

Richard Nixon helped seal the deal. Nixon had received nearly a third of the African-American vote in his unsuccessful 1960 bid for the White House, but when he ran and won in 1968 that share dropped to 15 percent. In 1972, he was reelected with just 13 percent of the Black vote. That was in part because the Republican brand was already tarnished among Blacks and in part because the Nixon campaign used the "southern strategy" to try to capitalize on racist white flight from the Democratic Party as more Blacks moved into it.

As Nixon's political strategist Kevin Phillips told the *New York Times Magazine* in 1970: "The more Negroes who register as Democrats in the South, the sooner the Negrophobe whites will quit the Democrats and become Republicans."

This may well be one of the greatest racial betrayals in American political history: the party to which Black people had pledged their devotion decided that it would be more fruitful to attract the people who hated them.

Practically, I suspect that for the foreseeable future, Black people will continue to vote overwhelmingly Democratic because of the Republican Party's countenance and even perpetuation of white supremacy and antiminority sentiments and policies. However, it is also easy to imagine a Republican Party that finds it can no longer win a national election without making drastic changes to appeal to a broader portion of the electorate. It is not impossible to imagine the party launching a remix of the southern strategy, this time to court not the Negrophobe, but the Negro.

I revel in this idea of forcing political parties to evolve. That is what political power looks like in action. And, this would be to the benefit of Black people. I personally find a functioning two-party system, or even a multiparty one, preferable to the single-party-dominated system that the failure of the Republican Party could yield. Multiparty competition is more likely to yield favorable policy outcomes not only for Black people but for all people, particularly if each makes a genuine effort to appeal to a broad coalition of voters.

It is my great hope that a mass movement south would so destroy the Republican Party's chances of winning the presidency or controlling the Senate that it would be forced to course-correct and become a desirable alternative for Black people and others. This is not without precedent: if Black people could eventually get over the Democrats' history of racism, they can eventually get over the Republicans'.

This could create competition for the Black vote, rather than it being what Princeton political scientist Paul Frymer calls a "captured constituency," voting overwhelmingly for a Democratic Party that too often takes its support for granted and ignores, rebuffs, or gives short shrift to its concerns, without much consequence. It is a simple truism: when people are

THE PROPOSITION

forced to compete for your vote, they play closer attention to your interests.

As James Baldwin wrote in his essay "Journey to Atlanta," "'Our people' have functioned in this country for nearly a century as political weapons, the trump card up the enemies' sleeve; anything promised Negroes at election time is also a threat levelled at the opposition; in the struggle for mastery the Negro is the pawn."

While politicians, particularly presidential ones, seek to entice fickle white votes, they simply seek to excite Black ones. They target them with policy and us with passion, a head-versus-heart strategy that yields a reality of wallet and well-being on one side and showy practices and shallow promises on the other. A return to the South in numbers capable of making Blacks able to deliver a state or to run it would permanently alter this dynamic.

As I've said earlier, in no way will I excuse the South for its atrocities. There exists in the region an architecture of endemic racial subjugation that its denizens refuse to relinquish and that must be pried loose.

This is instead to say that racism is everywhere in this country, just different shades, all cousins. And it is the hypocrisy of the northern and western cities—places practicing their own forms of corrosive oppression—that chafes most. White moderates in the North and West plead their desire to be better allies, but they refuse to recognize that the system uses their very presence as a weapon, employing brutal police tactics in the name of keeping them safe, promoting hypersegregation and imbalance in the name of keeping them comfortable. They want to "do something" but relinquish nothing. It's all an enormous pageant of faux probity in $80 yoga pants, holding $8 lattes.

They award themselves laurels for doing the least bit of labor to lessen the pain of an affliction rather than cure it. They focus on mitigating the impact of white supremacy rather than eradicating it. And, in its worst form, this misdirected, milquetoast nursing and nannying tries to fix people—Black people—with whom there is nothing wrong other than being trapped in wastelands of despair, starved of access, opportunity, and resources.

Among the wealthy who are thus inclined, they give a little and expect to be praised a lot. Philanthropy serves as a badge of benevolence, a kind of shield to justify excess, to enjoy privilege and the spoils of a system of white supremacy. Even more manipulative and self-serving, it offers a way of gaining cultural currency among the elite, to move up the social ladder from nouveau riche and nearer to old money. It's like tossing coins into the ocean for the joy of wishing, with the added benefit of a tax break.

In a perfect world the composition of community would be meaningless when aiming to ensure equity and equality. But we don't inhabit that world. In this world, America has had four hundred years to get right by Black people, and it has failed. There have been undeniable advances, but when barbarism is the base, the only way to move is up.

And, I must confess that at a certain point, I simply believe

THE PROPOSITION

that it is spiritually healthier to be in spaces, to create spaces, where you are wanted, honored, and loved, rather than ones where you are simply tolerated at best, or, worse yet, despised. Integration has its virtues, but it can also inflict spiritual and psychological violence. Nine experiences may be lovely, or at least tolerable, but the tenth is terrible. The mind settles on that tenth, using an extraordinary amount of energy to anticipate it, confront it, and overcome it. The unfortunate fraction becomes a consuming detriment.

I see Black students at elite, mostly white colleges in the North and West protesting poor treatment and demanding "safe spaces," and it is foreign to me. And it pains me. Some part of me always wants to say to them: You are spectacular. Being in your presence should be seen as an honor, not a burden.

I am taking an ever-dimmer view of granting the gift of my company to anyone who would reject it. As Zora Neale Hurston wrote in a letter to the *Orlando Sentinel* after the 1954 *Brown v. Board of Education* decision: "How much satisfaction can I get from a court order for somebody to associate with me who does not wish me near them?"

Hurston continued:

> *Since the days of the never-to-be-sufficiently deplored Reconstruction, there has been current the belief that there is no great[er] delight to Negroes than physical association with whites. The doctrine of the white mare. Those familiar with the habits of mules are aware that any mule, if not restrained, will automatically follow a white mare.*

I'm not one of those mules.

I went to an HBCU in the South not because I had to but because I wanted to. Before that, I went to an all-Black high school.

During my schooling, all of my spaces were safe. Almost every classroom I ever walked into, a Black person was the smartest person in that room, so as an adult I continued to believe that into every room I walked I could be the smartest person there. The security and comfort of being inculcated and enveloped by Black culture, not just becoming "woke" to it as intellectual pursuit, a form of activism, or need for demonstrative rebellion, produced in me an overwhelming confidence and also an ease of spirit.

Over the years I have found that many Black people with far more distinguished academic pedigrees than me also carry far more racial insecurity. It stymies creativity and growth. I liken this phenomenon to being covered with scars—scars from hand-to-hand combat with white racism because you are always outnumbered, always on the defensive. Black creatives keep issuing plaintive missives—movies and books, television offerings and an endless stream of podcasts—that feel to me to be a truly northern and western depiction of the Black experience. Many southern Blacks like myself have given up trying to persuade and explain. (Although I lived in New York most of my life, my soul has always been in the South, to which I have now returned.)

As the great Toni Morrison once put it:

THE PROPOSITION

The function, the very serious function of racism is distraction. It keeps you from doing your work. It keeps you explaining, over and over again, your reason for being.

Time, energy, and passion are limited commodities in a life. Every minute I spend trying to fix the flaw in another is a minute I take away from loving my family, being in my community, and doing my work. I refuse to give racism my minutes. Furthermore, it is outrageous to even expect the oppressed to heal the oppressor. That, in fact, is another form of oppression.

THREE

THE PUSH

● ● ●

IF A WHITE MAN WANTS TO LYNCH ME, THAT'S HIS PROBLEM. IF HE'S GOT THE POWER TO LYNCH ME, THAT'S MY PROBLEM. RACISM IS NOT A QUESTION OF ATTITUDE; IT'S A QUESTION OF POWER.

—Stokely Carmichael,
speech at the University of California, Berkeley

I could hear a quiver in the voice on the phone, a voice that was soft, shaky, shaken, unsure of itself.

It was my son, but not my son. This was not the Tahj Ali I knew. This young man was strangely alien to me, inhabited and reduced, consumed by a fear and tension that neither he recognized in himself nor I in him.

A police officer had detained him and pulled a gun as my son was exiting the library at Yale.

I was late for a dinner with friends near Lincoln Center. I was circling the block in search of a place to park when the call came. I answered with the car's speaker system. I knew right away that something was off.

"What's wrong?" I asked. And so he began to recount his terror.

As soon as I heard the words "pulled a gun," half the blood and half the life seemed to drain out of me. I immediately went cold and tingly and stiff.

I pulled the car over and picked up the handset. Somehow this was better. I needed him, or at least his voice, as close to me as possible, pressing against me, not floating coldly in the air.

Before he could continue, I interrupted with the first question any parent thinks to ask: "Are you okay?"

He said he was. But, in truth, he was not.

He continued with his tale. It had been a busier-than-usual Saturday for him. His girlfriend, a pre-law major at Cornell, was visiting. My son was a third-year evolutionary biology major planning to go to medical school and become a doctor.

It was January 24, 2015.

That weekend he was rushing to produce a promotional video for his fraternity. To do so, he had checked out a video camera, a microphone, and a tripod from the campus library, where he also worked. But, one could keep the equipment for only a few days at a time, so every three days he would return it and check it out again.

That is what he was doing that evening.

As he was leaving the library by an underground passage he flung the equipment's straps over his shoulder, braced for the cold, and ascended the stairs. It was freezing. He was wearing one of my jackets I'd given him, a navy blue peacoat, a sweater cap—blue with a rust-colored ring—and hiking boots. He was a true Brooklyn boy, partial to Timberlands or any boots that looked like them. His attire wasn't quite suitable for the weather, but it was serviceable.

When walking up the stairs he saw an officer running down one of the cross-campus paths near him, which he thought unusual, but it didn't much faze him. He turned to take the sidewalk back toward his dorm.

He glanced over his shoulder to see that the officer had stopped and was now staring in the direction he was walking. Still, nothing registered with Tahj as problematic. Why should it have? He kept the casual but purposeful gait that cold weather compels, the same as the rest of the people on the sidewalk.

Then he heard a voice yell, realized that the voice was yelling at him, and he turned to see the officer remove his gun from its holster. As the gun began to move, everything in my son sank. He instinctively lowered his head, and raised his hands. He fell to his knees and his body collapsed chest-first on cold concrete. Panic pulled him down. If the sidewalk could have opened it would have swallowed him. It was a Black man's ancestral muscle memory of submission, of surrendering for survival.

I knew this instinct well, this horror stored in the blood. It is the same one that had seized me one spring night at Grambling State University some twenty-seven years earlier. A police officer in the mostly white neighboring town of Ruston, Louisiana, had trailed a car my friend was driving and in which I occupied the passenger seat. Just before we were about to cross the city limit and leave his jurisdiction, he turned on his flashers and pulled us over. He told my friend that he had made a turn without signaling—which was a lie because we hadn't made any turns—and demanded to see my friend's license, insurance, and registration.

My friend motioned toward the glove box in front of me, and as I opened it to retrieve the documents a switchblade comb tumbled out. It was the kind of toy you won at the parish fair when you had the skill to steady the tiny crane claw in the glass enclosure so that you could pick up an item and swing it to the prize chute.

As the comb tumbled out, the officer drew his gun. I assumed that it was trained on me, but I couldn't be sure, no more than my son could be sure that night at Yale. My hands went up instinctively, my body stiffened, and my gaze locked, not on anything really, just forward, away from the weapon.

When my friend protested the officer's treatment, the man replied flatly that he could make us lie down in the middle of the road and shoot us in the back of the head, and no one would say anything about it.

This threat of execution undid us. How could this be? We were the smart kids, the good kids, scholarship kids. I had been the freshman class president and that semester I would be elected sophomore class president and my friend would be elected vice president. And yet, there we were, on the side of a dark road, on the wrong end of a loaded gun.

I have never forgotten the fear and helplessness and rage of that moment, and it kept playing in my mind as my son spoke. Him lying on the ground where an officer had threatened to put me, his body splayed, stomach to pavement, face a nose away from it, another generation of Black man being inducted into "the club," the ignominious order to which so many Black men belong: the harassed and detained, the roughed up and beat down, the killed and the clinging to life.

In 2020 a Kaiser Family Foundation poll asked respondents if they had ever been stopped or detained by the police because of their race or ethnicity. Forty-one percent of Black people said yes. Only 5 percent of white people said the same.

These stops have become a rite of passage for Black men, a bonding in trauma, a ritual we try our best to laugh off at the barbershop or bar.

THE PUSH

Somehow, I thought, naively, that I could save my boys from their initiations, that I could mock fate and skirt the rule. I moved out of the South, to the supposedly liberal Northeast. I moved to one of the best elementary school districts in Brooklyn when they were younger and sent them to a top private school, the bucolic Riverdale Country School, when they were older. I stayed on top of them, making sure that they behaved and excelled, and they stayed out of trouble and on track while bristling at the pressure I applied.

Tahj had studied Mandarin while in high school and traveled to China as a high school junior to teach English and immerse himself in Chinese culture. He applied to twelve colleges—the Black boy with great grades, high test scores, and impressive extracurricular profile who thought at the time he wanted to study an Eastern language—and was accepted to eleven. Yale, Princeton, and the University of Pennsylvania were among them, but also historically Black schools like Morehouse in Atlanta and Howard in DC.

I kept all the acceptance letters. None of my grandparents had gone to college. My mother's parents, the only grandparents I remember before they died, worked their whole lives with their bodies until their bodies gave out, my grandpa Fred on railroads and my grandma Mama Tat cleaning white women's houses and comforting white women's babies. My father, a former bandleader turned construction worker, had never been to college. My mother finished her studies only when I was already five years old; I was the youngest of five boys. Tahj's success, I thought, was an achievement for lineage and legacy, an achievement for the great-great-grandmother, born a slave, whose picture I kept in a photo album and often pondered—the hard look of pain and trauma on her face, no smile attempted or managed, hair wild and standing like a sculpture—concluding that her worries were by no means about the shallowness of glamour but the existential threats to her survival.

My son chose Yale, after a school visit during which they partied almost nonstop as best I could tell. He would be the first person to attend an Ivy League school in our family, and I took a particular, oddly elitist pride in that, one that I would later come to deeply question.

All that grooming and performing, all that excelling and sacrifice in the Northeast led him to the same fate to which my desperation and scraping had led me in the Deep South.

He "matched the description" of an alleged criminal. That was the pretext. There was apparently an intruder on campus that night entering students' rooms. As Yale explained:

> The intruder was described by students as an "extraordinarily tall black male" wearing a "black coat, red and white beanie cap" and with "orange details" on his shoes.

My son is six one, and his clothing wasn't an exact match; the colors of the items were all off. But, of course, he was Black, and so out came the gun.

My son said that what he felt at that moment was "flat," everything pushed down and squeezed out. Not anxious or petrified as he thought he might have been or should have been, but numb, his mind having moved to a placid place, a plane on which everything around him darkened and fell away, one on which the only things that now existed were the barking pulsing down on his head from the man with the gun. He concentrated intensely, exclusively, on the tone and content of his own responses.

And that was the problem: the gun. I didn't really object to his being stopped and questioned. That campus was his home too, and he and I both wanted it kept as safe as possible, for everyone, including him. But, it was clear to me that it was the rarity of his Blackness in this white space that made the lack of matching descriptor to the actual suspect inconsequential.

The presence of the gun shifted the dynamic. I knew well from the reporting for my columns in the *New York Times* that a drawn gun was only one flash of fear, one misunderstood flinch, one disobeyed order from the ending of a Black man's life.

Two weeks before my son was detained, I had written about the Cleveland killing of twelve-year-old Tamir Rice, a boy playing alone with a toy gun in a park. It was broad daylight in an open-carry state.

Officers had responded to a 911 call of a "guy" pointing a gun. The caller said the gun was "probably fake," but that fact wasn't conveyed to the officers.

One of the officers was Timothy Loehmann, a twenty-six-year-old just months on the job who had been fired from his last position because his superiors thought him unfit and unstable. He shot Rice within 1.5 seconds of arriving in the park, the cruiser barreling across the grass to reach the boy, and improbably the officer asserted that he ordered Rice to put his hands up three times before he shot. (Coincidentally, the officer at Yale also said he called out to my son three times before my son responded. If that happened, the first two commands were not heard by my son or didn't register with him.)

Tamir's fourteen-year-old sister came running, screaming, to his aid from a nearby building. But the officers tackled her, shackled her, and shoved her into the back of their car. Tamir was still alive. He was just feet away. But she couldn't reach him. She couldn't comfort him. All she could do was scream.

Tamir's mother, Samaria, arrived from the other direction, only to discover the ghastly scene: one child sprawled out with a bullet in his stomach and another locked behind glass, screaming hysterically.

As Samaria told me: "At that point, I went into shock, because at that point I'm trying to figure out: 'What is going on? What happened? What did he do?' In my head it's like, 'What did he do bad enough for you guys to shoot him?'" In the scrum, her oldest son was also detained by the officers.

She had to make an impossible decision: Go in the ambulance with the child who got shot or stay and try to free the other two who were now in the custody of the man who shot him?

Samaria's pain and loss were palpable. I felt her words in the most intimate of ways, like I too knew Tamir and lost him. She was angry, fuming, and she had every right to be. Tamir's death had opened a hole in her that could never be filled. I never want to be Samaria. No one does.

But it was another opinion piece that appeared in the *New*

THE PUSH

York Times the day before my column about Tamir and his mother was published that resonated with the greatest depth. It was written by Isabel Wilkerson, the author of *The Warmth of Other Suns: The Epic Story of America's Great Migration.*

In it, she crystalized something that I had picked up on after years of reporting on the police killings of Black men and women: many of the highest-profile cases had taken place in the North, West, and Midwest in destination cities, "the place to which generations of African-Americans fled to escape the state-sanctioned violence" in the South.

This is not a perfect construct, as there are notable examples of extrajudicial police killings in the South as well—Alton Sterling in Baton Rouge, Walter Scott in North Charleston, Eric Harris in Tulsa, Rayshard Brooks in Atlanta—but for me there is a particular resonance in Wilkerson's reasoning.

Police killings of unarmed Black people are in fact surprisingly evenly distributed. The *Washington Post* tracked 124 killings of unarmed Black people from January 2015 to August 2020. The South has 55 percent of the Black population and 62 percent of the killings. The Northeast had 17 percent of the population, but 12 percent of the killings. For the Midwest it was 18 percent of the population and 16 percent of the killings. And for the West, the region that overindexed to the greatest degree, the numbers were 10 percent of the population and 19 percent of the killings.

And yet, you need to look more closely at the breakdown of states to get the full picture. The classification of the killing of people like Breonna Taylor, murdered in Louisville, Kentucky, for example, isn't so straightforward. Kentucky is technically a southern state, but Louisville was also a destination city, particularly in the second wave of the Great Migration. All the states in the upper South—Kentucky, West Virginia, Virginia, Maryland, and Delaware—were destination states, as was the District of Columbia.

When those states are excluded from the South's tally, the region actually underindexes on its population-to-killings ratio. It should also be noted that just two of the South's total of sixteen states, with no exclusion—Texas and Florida, states not among my recommendation for reverse migration—are home to a staggering 43 percent of the region's killings of unarmed Black people.

One thing that the disastrous 1994 Violent Crime Control and Law Enforcement Act got right was that it empowered the Department of Justice to sue police departments that "engage in a pattern or practice of conduct" of civil rights violations. The DOJ often enters with such police departments into court-enforced agreements called consent decrees until the problems are judged to have been remedied. As of late 2020, there were thirteen police departments in America operating under consent decrees, most of them in destination cities, including Detroit, Los Angeles, and Portland. Only two of the agreements are in the South: New Orleans and Meridian, Mississippi.

Of the sixteen or so consent decrees entered into since the 1990s but since having been terminated, most are in destination cities in the North and West. These are cities like Los Angeles, Detroit, Pittsburgh, Buffalo, Cincinnati, and Hyde Park, Illinois.

I believe that what we're witnessing now is a transference of terror: you can draw a distinct line between the lynching

THE PUSH

that Blacks in the South fled during the Great Migration and the oppressive policing, including the shootings of unarmed Black men, that we see today, many in the destination cities.

Eric Garner was choked to death in New York.

Freddie Gray died of a severed spine in the back of a police van in Baltimore.

Sam DuBose was shot through the head during a traffic stop in Cincinnati.

Philando Castile was shot in a car with his girlfriend and daughter near Saint Paul.

Michael Brown was killed in Ferguson, outside Saint Louis.

Laquan McDonald was gunned down in the street in Chicago.

Stephon Clark was killed in his grandmother's backyard in Sacramento.

George Floyd Jr., in the street with a knee on his neck in Minneapolis.

These are many of the same cities that experienced the violent conflict of race riots in the late 1910s, and these are the same cities that experienced race riots some fifty years on during the 1960s, giving rise to a great fear in Martin Luther King and occupying much of his time there. There is a semicentennial repetition to these spikes in racial violence, and the revolts against that violence.

Indeed, not much has changed in the trajectory of suffering for black people living in these cities. Months after the Watts riots of 1965, King confessed in an essay written for the *Saturday Review* that the civil rights movement had been a regional movement, centered in the South, with the South reaping the lion's share of the rewards. But he noted:

In the North, on the other hand, the Negro's repellent slum life was altered not for the better but for the worse. Oppression in the ghettos intensified. To the homes of ten years ago, squalid then, were added ten years of decay. School segregation did not abate but increased. Above all, unemployment for Negroes swelled and remained unaffected by general economic expansion. As the nation, Negro and white, trembled with outrage at police brutality in the South, police misconduct in the North was rationalized, tolerated, and usually denied.

Much of this continues to be true some fifty-five years on.

Specifically, the recent police shootings aren't happening in Selma, or Birmingham, or Montgomery, or Greensboro, or Memphis, or Little Rock. In a way, the urban North and West have become the New Civil Rights battleground, with its own movement in the form of Black Lives Matter, and yet these cities have been slow to come to terms with the viciousness and persistence of their own brand of racism.

This new battle is more furtive, and therefore harder to fight and win, because the most effective forms of racism today aren't shouted out loud or written explicitly into ordinance. Systems now do the bulk of the work; there is a perpetualness to racialized poverty and oppression. At a certain point—one long since passed in America—little effort is required to maintain the structures. Hopelessness and despair seep into the psyche. The damage becomes generational inheritance and culture caste.

And while the scale of these killings comes nowhere close to the scale—and savagery—of the lynchings of the past, modern technology and ease of distribution mean that the killings we see on television, on the internet, and in printed publications reach exponentially more people than one could even have imagined at the turn of the twentieth century.

As Bryan Stevenson of the Equal Justice Initiative explains, when lynching stopped, legal executions picked up. Essentially "the killings moved indoors": the mobs gave way to the magistrates. But now, those executions have moved into the virtual space. We have access to unlimited images of Black death on the phones in our pockets. State violence performed as street violence became streamed violence.

The use of imagery to disseminate, normalize, and institutionalize terror is a tool of antebellum vintage. White supremacists printed photographs of lynchings on postcards, the dangling, sometimes charred bodies symbols of perverse racial pride and fierce dedication to racial protections and purity. These postcards not only commemorated terror, they rendered it quaint.

But one of the first images to move Black America to a concerted national outrage resulting in orchestrated action was the mutilated face of Emmett Till.

Till was a Black fourteen-year-old Chicago boy who was visiting his great-uncle in Mississippi during the summer of 1955. The story went that the boy said something to, and whistled at, a white woman, Carolyn Bryant Donham. This was a line not crossed in those parts. In the wee hours of the night, two white men—one Carolyn's husband, Roy, the other a former soldier called "Big" Milam thought by his associate to have "handled negroes better than anyone in the country"—kidnapped Emmett from his family's home and mercilessly pistol-whipped him. After he refused to be cowed or broken, saying by one account, "You bastards, I'm not afraid of you. I'm just as good as you," they took him to the banks of the Tallahatchie River, where Big Milam shot him in the head, then the two men tied the metal fan of a cotton gin around his neck with barbed wire and dumped his body in the water.

When Emmett's body was fished from the river three days later, it had already begun to decompose. He was unrecognizable. His body was identified because he was wearing a ring that had belonged to his father.

His corpse was sent back to Chicago for burial. His mother, Mamie, collapsed at the sight of the casket, just two weeks after she had kissed her son goodbye on his trip.

She insisted that the casket be opened. As she recalled: "I saw that his tongue was choked out. I noticed that the right eye was lying on midway his cheek. I noticed that his nose had been broken like somebody took a meat chopper and chopped his nose in several places. As I kept looking, I saw a hole, which I presumed was a bullet hole, and I could look through that hole and see daylight on the other side. And I wondered: Was it necessary to shoot him?"

Mamie required the world to gaze upon Emmett's face, forcing everyone to see what had been done to her baby.

Tens of thousands of people filed past Emmett's glass-covered casket during his wake to gaze at the contorted mass of flesh that was once his face. As many as one hundred thousand viewed his body before his funeral.

THE PUSH

A little over two weeks after Emmett was buried, the men who killed him were acquitted, after only sixty-seven minutes of jury deliberations. One juror is said to have told a reporter that the deliberations wouldn't have taken that long if the jurors hadn't taken a break to drink a pop.

After the acquittal the killers kissed their wives, lit cigars, and posed for pictures. A few months later, in January 1956, *Look* magazine would publish what it called "the real story of that killing—the story no jury heard and no newspaper reader saw." In it, Till's killers appeared to confess.

As *Look* quoted Big Milam about the moment he decided to kill Emmet Till:

Well, what else could we do? He was hopeless. I'm no bully; I never hurt a nigger in my life. I like niggers—in their place—I know how to work 'em. But I just decided it was time a few people got put on notice. As long as I live and can do anything about it, niggers are gonna stay in their place. Niggers ain't gonna vote where I live. If they did, they'd control the government. They ain't gonna go to school with my kids. And when a nigger gets close to mentioning sex with a white woman, he's tired o' livin'.

Black people not "staying in their place," the idea of white displacement, of miscegenation and the corrupting of "white purity," the idea of Black governmental control and the possibility of Black vengeance, have long been the source of white terror in this country. It has motivated white violence and white policy.

(Some six decades later, Donham would reportedly tell the author Timothy B. Tyson that she had lied about Emmett making any crude gestures toward her.)

Of all the deaths and all the images of the twentieth century, it was the picture of Till that proved most potent. *Jet* magazine published a photograph of Mamie Till, being braced and held up by a man as her eyes were cast down at Emmett's battered, bulbous, unrecognizable face. *Time* magazine would label this image one of the one hundred most influential pictures ever taken.

Although, as law professors Margaret A. Burnham and Margaret M. Russell have argued, there were hundreds of "disappeared" Black people in this country "who were victims of racial violence from 1930 to 1960," Till became the most pivotal. Jesse Jackson is credited with designating Emmett's murder the "Big Bang" of the civil rights movement. University of Illinois professor Christopher Benson, coauthor of the 2003 book *Death of Innocence*, has called Till's death the "first 'Black Lives Matter' story." His death was immeasurable in its effect on young Black people at the time. It mobilized them in a way that wouldn't be seen again until the killing of seventeen-year-old Trayvon Martin by a neighborhood vigilante, George Zimmerman, in 2012, and then again with the killing of George Floyd in 2020.

Acts of terror took on another dimension when recorded on video, beginning with Bull Connor's barbarous use of dogs and fire hoses to squash protests among Black youth in Birmingham in the early 1960s. But then the geography shifted.

THE PUSH

The current era, I would argue, was set off and framed in 1991 when Los Angeles Police Department officers mercilessly beat motorist Rodney King on the side of the road.

The beating was caught on video by a private citizen named George Holliday, who filmed the carnage from his balcony with a Sony Handycam. It showed "officers taking turns swinging their nightsticks like baseball bats at the man and kicking him in the head as he lay on the ground early Sunday," as the *New York Times* described it.

I was a twenty-year-old college student at the time, and like many of my contemporaries, the repeated airing of the footage altered something in me, ignited something. This was two years after my friend and I had been pulled over by the country cop and threatened with execution. But somehow, in my mind, that white man standing in the darkness had been an aberration, an anomaly. This thing, this orgy of rage and terror being beamed into my living room, was pack behavior. This was an unleashing of violent cultural ethos.

King's injuries included a "fractured cheekbone, 11 broken bones at the base of his skull, and a broken leg," according to the *Los Angeles Times*.

Earlier in the day, before the beating, one of the officers who participated had typed a message on a computer terminal in a squad car, referring to a domestic dispute among Blacks this way: "Sounds almost as exciting as our last call. It was right out of 'Gorillas in the Mist.'"

Like many people at the time, I thought that justice was assured because the beating had been captured on video. I was wrong. The four officers were acquitted of assault. Those acquittals sparked the Los Angeles riots of 1992, which when judged by estimated insurance losses were by far the most costly riots in American history, even when adjusted for inflation. The total cost was nearly $1.4 billion in 2017 dollars—nearly as much as the next nine most costly riots combined.

Small wonder then that according to the *LA Weekly*, "by the early 1990s the LAPD was viewed by many African Americans in South L.A. as the occupying army of a hostile government."

The years that followed saw the advent of cable news, camera phones, body cameras, and social media. Now, everyone could capture video and everyone could share it. These images would provide much of the fuel for Black Lives Matter, provide documentation of what Black people had long bemoaned, often to white people's disbelief: that their very own Officer Friendly was often the Black community's abusive overseer.

But, the incessant airing of—and unlimited access to—these violent videos has led to the creation of a kind of functional cinematic genre, part social horror, part morbid pornography. And in so doing, they create a tableau of collective Black terror, a state of perpetual Black trauma.

As Monnica T. Williams, a psychologist and associate professor at the University of Connecticut, wrote in *Psychology Today* in 2015:

> We are surrounded by constant reminders that race-related danger can occur at any time, anywhere, to anyone. We might see clips on the nightly news featuring unarmed African Americans being killed on the street, in a holding cell, or even in a church. Learning of these events brings up an array of painful

THE PUSH

racially-charged memories, and what has been termed "vicarious traumatization." Even if the specific tragic news item has never happened to us directly, we may have had parents or aunts who have had similar experiences, or we know people in our community who have, and their stories have been passed down. Over the centuries the Black community has developed a cultural knowledge of these sorts of horrific events, which then primes us for traumatization when we hear about yet another act of violence. Another unarmed Black man has been shot by police in our communities and nowhere feels safe.

Too many of these "vicarious traumatizations" are emanating from destination cities. These cities no longer feel safe, if they ever truly did.

The results of a weekly survey by the US Census Bureau found in 2020 that, according to the *Washington Post*, "the rate of black Americans showing clinically significant signs of anxiety or depressive disorders jumped from 36 percent to 41 percent in the week after the video of Floyd's death became public. That represents roughly 1.4 million more people."

These images expose the extreme of a sobering reality, the inevitable result of a pattern of betrayal on the issue of criminal justice—by liberals and conservatives alike—over the last few decades. The systemic brutality of overpolicing started to take on a more formal structure during and after the race riots of the 1960s, when police departments across the country began to establish SWAT teams. One leader in this effort was the destination city of Los Angeles.

Many looked to the King riots as the flash point in the escalation of police brutality against Black Americans. But I would argue the history is more complicated and that some of the villains aren't the expected ones. In 1988, a twenty-two-year-old rookie police officer named Edward Byrne was ambushed and killed on the orders of a drug lord while guarding a witness in Queens, New York. This was a true tragedy. More than ten thousand officers mourned him at his funeral.

That same year, George H. W. Bush was running his tough-on-crime presidential campaign against Michael S. Dukakis, and successfully using Willie Horton—a Black murderer who raped a white woman while on furlough from prison—as a weapon against his opponent.

A few weeks before the election, Bush made his first general election campaign appearance in New York City, surrounding himself with uniformed policemen and accepting Byrne's badge from his father. Bush said at the event that he wanted to use the occasion "to help define for you the man I am running against, throw a little red political meat out there."

Bush subsequently won the endorsement of the Fraternal Order of Police, and he carried forty states on Election Day, delivering a crushing blow to Dukakis and the Democrats. Bush would say later that he kept Byrne's badge in the Oval Office.

Also in 1988, the Edward Byrne Memorial Justice Assistance Grant program, named after the fallen officer, was established by the Anti-Drug Abuse Act to supercharge the war on drugs—a disastrous boondoggle that would devolve into a war waged largely against marijuana usage by Black and Hispanic men. As the American Civil Liberties Union pointed out in 2011, "The racial disparities are staggering: despite the fact that whites engage in drug offenses at a higher rate than African-Americans, African-Americans are incarcerated for drug offenses at a rate that is 10 times greater than that of whites."

Democrats, determined to never be outflanked on crime again, began to move dramatically to the right, in some cases further to the right than Republicans themselves.

In 1989, New York's mayor Ed Koch, a Democrat, spoke of Byrne's killing and those of other officers, ominously declaring, "The pendulum protecting those who violate the law has swung too far."

In 1992, the Fraternal Order of Police endorsed Bush over Bill Clinton, although Clinton gained the endorsements of two smaller police unions. But in 1994, Clinton championed a crime bill that would have the federal government pay for one hundred thousand additional police officers, a move the *Washington Post*'s Fact Checker blog has since asserted helped neutralize Republican "weak on crime" attacks against Democrats, and in 1996, securing for Clinton the first endorsement of a Democratic presidential candidate by the Fraternal Order of Police.

The group then endorsed John McCain over President Obama in 2008, but Obama was able to secure the endorsement of the National Association of Police Organizations. For one thing, Obama had promised that he would restore funding to the Byrne Grants program, which George W. Bush had sought to eliminate. Conservative groups backed Bush's proposal to eliminate the program, saying it "has proved to be an ineffective and inefficient use of resources."

In the last year of the Bush administration, the program's financing had been reduced to $170 million. In March of that year, fifty-six senators signed on to a "bipartisan" letter to ranking members of the Senate Appropriations Committee urging them to restore nearly $500 million to the program. Only fifteen Republicans signed the letter.

Some of this might have had to do with conservatives' deep contempt for unions, but that is another matter.

When accepting the endorsement from the National Association of Police Organizations, Obama issued a statement lauding the organization:

Day and night, America's police work tirelessly to make sure that our families are safe and our communities are strong. It's a basic responsibility to make sure that our officers have the support they need to fight crime and bring criminals to justice. Too often under the Bush Administration, we've failed to live up to that obligation.

Obama's running mate, Joe Biden, had been the chief author of the 1994 crime bill that not only flooded American streets with more police officers, but also expanded the death penalty, and increased funding for border patrols.

Obama followed through on his promise, as the 2009 stimulus package presented these Democrats with the opportunity, and they seized it. The legislation, designed by Democrats and signed by President Obama, included $2 billion for Byrne Grants to be awarded by the end of September 2010. That was nearly a twelve-fold increase in financing.

In 2012, the National Association of Police Organizations again endorsed the Obama-Biden ticket. Touting the administration's "unwavering support," they proclaimed, "There are simply no better friends of law enforcement—and no stronger choice to lead this nation for another term—than President Obama and Vice President Biden."

And yet Democrats soon realized that they were in bed with fickle partners. When protests erupted after grand juries did not indict police officers in the deaths of Michael Brown and Eric Garner in 2014, the president, the attorney general, and the mayor of New York weighed in with how they personally saw tensions between the police and communities of color, how those tensions had affected their lives and their understanding of the protesters' feelings.

In a statement, Eric Holder, the US attorney general and a New York native, who had revealed the humiliation of his own unfair stops by police, explained:

We have all seen the video of Mr. Garner's arrest. His death, of course, was a tragedy. All lives must be valued. Mr. Garner's death is one of several recent incidents across the country that have tested the sense of trust that must exist between law enforcement and the communities they are charged to serve and protect. This is not a New York issue or a Ferguson issue alone. Those who have protested peacefully across our great nation following the grand jury's decision in Ferguson have made that clear.

When two New York City police officers were ambushed in Brooklyn and shot at point-blank range by a Baltimore man, apparently angry over the deaths of Brown and Garner and intent on killing officers, no time was wasted in attempts to tie all these political figures to the killings. Former

New York City mayor Rudy Giuliani even said, "We've had four months of propaganda, starting with the president, that everybody should hate the police." This was a lie.

And the National Association of Police Organizations, which had so effusively praised Obama and Biden, chimed in:

> *Politicians have created an environment of extreme hostility in communities across the nation. Our nation's leaders continue to crucify and demonize law enforcement officers as these officers work tirelessly and selflessly to protect us. When will this end? When will our leaders work with us, not against us, to build trust between officers and the communities that they serve?*

Whatever the merits of programs like Byrne Grants, they are outweighed by the damage being done. Financing prevention is fine. Financing a race-based arrest epidemic is not. This is exactly what we have witnessed for decades.

And all the while white people in destination cities, in particular, either encouraged or turned a blind eye to the unconscionable, oppressive excesses by their own police departments because, whether they admit it or not, they are committed to the same control over the Black body to which the law has been dedicated in this country from the beginning, a strategy that the modern North has adapted from the historical South.

The same year that the Civil War ended, the two states among those with the largest percentage of Black people—Mississippi, with 54 percent in the 1870 census, and South

Carolina, with 59 percent—passed what came to be known as "Black codes," a repressive slate of laws to "regulate the Domestic Relations of Persons of Colour." These laws forced freedmen into contractual labor agreements, which looked eerily similar to slavery, with white farmers. The South Carolina act even stipulated that "all persons of color who make contracts for service or labor, shall be known as servants, and those with whom they contract, shall be known as masters." Freedmen without "some lawful and respectable employment" could be charged with vagrancy. They literally made Black unemployment a crime for Black people.

Other provisions included, "No person of color shall migrate into and reside in this State unless, within twenty days after his arrival within the same, he shall enter into a bond, with two freeholders as sureties," and "Servants shall not be absent from the premises without the permission of the master."

Perhaps more important, the acts restricted gun ownership and allowed for the seizure of weapons "unlawfully kept." White people were petrified by the thought of an armed Black populace, one that outnumbered them in South Carolina's case, knowing well the horrors they had visited upon the flesh of those who would carry those firearms.

This was the consuming fear of revolt and retribution. This was a white terror.

This particular terror has long been the fire fueling white vigilantism in America.

Dylann Storm Roof, a sheepish, boy-looking man with cowlick hair, showed up at Charleston, South Carolina's historic Emanuel African Methodist Episcopal Church in June 2015 asking for the minister. The Bible study group invited him in. He sat with them for an hour. As the parishioners lowered their heads in prayer, Roof rose and started spraying the room with bullets. During the shooting, twenty-five-year-old Tywanza Sanders, already injured, forced himself to his feet and asked, "Why are you doing this?" Roof responded, "I have to do this. You are raping our women and you are taking over the world." When the shooting was over, nine Black bodies lay lifeless, including Sanders's. Six of the dead were women, three were men. Roof was there, in part, to defend the purity and sanctity of white femininity.

This violent defense of white femininity has a long history.

It was the putative reason for the horrific lynching of Jesse Washington in 1916 in Waco, Texas. Washington was accused of raping and murdering the wife of his white employer. The *Chicago Defender* printed a letter from a white resident of Waco who had witnessed the lynching. As *The Atlantic* summarized the letter's contents in 2016:

> *A mob of "fifteen to twenty-thousand men and women intermingled with children and babies in their arms" gathered to torture Washington and then burn him at the stake. Accused of the murder of a white woman several miles from his home, Washington was convicted by a jury despite scant evidence. Then, as happened all too often, Washington was dragged from the courtroom, hung from a tree, and burned on a funeral pyre. "The crowd was made up of some of the supposed best citizens of the South," the letter writer noted. "Doctors, lawyers, business men and Christians (posing as such, however). After the fire subsided, the mob was not satisfied: They*

THE PUSH

hacked with pen knives the fingers, the toes, and pieces of flesh from the body, carrying them as souvenirs to their automobiles." The correspondent went on to conclude that it was absurd to send soldiers to Mexico "when the troops are needed right here in the South."

The defense of white femininity was the reason for what came to be known as the Tulsa Riots of 1921, which erupted after a nineteen-year-old Black boy was accused of assaulting a seventeen-year-old white girl on an elevator. When the riots were over, as many as three hundred people had been killed and eight thousand left homeless. As 103-year-old Olivia J. Hooker, one of the last known survivors of the riot, recounted in 2018 as I sat with her in her White Plains, New York, home, "It took years for me to get over the shock of seeing people be so horrible to people who had done them no wrong."

It was the reason fourteen-year-old George Stinney Jr. was sent to a South Carolina electric chair in 1944.

Stinney was accused of murdering two white girls. He was tried just a month after his arrest. The same day that the all-white jury was empaneled, the trial commenced. It lasted only a few hours. Stinney's assigned defense lawyer, a white man, called no witnesses and cross-examined none. The jury deliberated for ten minutes. Guilty. The boy was sentenced to death by electrocution, which took place just two months later. There were no appeals or requests for a stay.

On execution day, the guards had a hard time fitting Stinney's tiny five-foot, one-inch, ninety-five-pound frame into the chair. They had to sit him on a book in the chair. Some say it was a phone book; others say it was the Bible. The 2,400-volt blast of electricity caused the death mask to slide off his face.

Seventy years later, in 2014, South Carolina threw out Stinney's conviction.

The list of Black men lynched, and Black communities terrorized, under the guise of defending white women is perverse in its length.

White women have known from the beginning in this country that they possess this power, the power to activate white supremacy and spur it to extreme violence. They merely have to feign victimhood, to appear afraid and affronted, to muster a tear or a scream. The activation of white terror is a white woman's soft power.

We like to masculinize white supremacy, to presume it reeks of testosterone, when in fact, it is just as likely to be spritzed by perfume. Forty percent of slave owners were white women. It was white women who made the market for Black women's breast milk and who were attended by Black women in the big house. It was white women who upheld much of the day-to-day white supremacy—the schoolteachers, the store clerks, the waitresses. And it is now often white women activating police interactions with Black people.

The geography of this exercise of that power has shifted from the South to the destination cities of the North and West, as white women call the police to "report" Black people for doing the mundane, everyday things that people do, essentially for "living while Black." It manifests in a white woman calling the police on Harvard professor Skip Gates entering his own home in Cambridge, Massachusetts. It manifests in a white woman calling the police on Black people having a cookout in an Oakland park. It manifests in a white woman calling the police on a Black child selling bottled water near her home in San Francisco.

The list of these outrages is nauseatingly long, but they point to the way white people, particularly white women, in these supposedly liberal destination cities, with the ease of 911, have warmed to a new way to use the police—still often an overwhelmingly white power—as their muscle, as a cudgel against Black people who stir within them even the slightest unease. The police have become instruments of their intolerance, a way to swing the ax without using one's hands. There is no longer a need to yell or taunt or spit. One need only dial. Keypad tones are the new barking hounds. White wrath can hide behind a phone call and a blue wall. These women now even have their own pejorative, Karen, personified by the woman who called the police on Christian Cooper, a Black birdwatcher in Central Park.

It is all about white power over Black bodies.

Dr. King called the riots in destination cities of the 1960s "the language of the unheard." In that language is written the repercussions of economic oppression and economic allocation that inform racial inequality and that challenge America's commitment to equality. But when white people heard that language, they ran. During the 1960s, '70s, and '80s, white flight depopulated destination cities and cratered their tax bases. I saw this firsthand when I moved to Detroit in the early 1990s. The city had been hollowed in the center. The joke there at the time was that you couldn't buy a refrigerator inside the city limit. But it wasn't a joke. There wasn't a single appliance store in Detroit, then the seventh-largest city in the country.

Whenever Black people make progress, white people feel threatened and respond forcefully.

Emancipation and the Civil War gave rise to the Ku Klux Klan, which formed just months after the war ended. The Supreme Court's decision in *Brown v. Board of Education*, striking down racial segregation in schools, gave rise to the white supremacist Citizens' Councils. The election of the first Black president gave rise to the Tea Party, which was formed soon after Barack Obama was sworn in.

It took centuries for America to hone its instruments of oppression. Every time part of it fell, it simply reemerged in a more elegant form. Battling racism in this country is like cutting heads off the Hydra.

Police aggression against Black citizens—and white people's willful blindness to it—is one of the latest incarnations, and that aggression is linked to white urbanization and gentrification.

White people, particularly white millennials, are having their own Great Migration as they flood out of small towns and into big cities. In many cases, this reverses decades of white flight from those cities. This creates housing pressure and higher prices. These young newcomers are willing and able to pay a premium and to move into areas long inhabited by Black and brown people. In some cases this forces out the established residents. Researchers have found that nearly 110,000 Black people were displaced from gentrifying neighborhoods from 2000 to 2013, and most of those displacements took place in destination cities.

For these young, college-educated white people—the so-called creative class—their presence is weaponized as a rationale for Black removal; their gentrification is the justification.

And on the fringes, often, of destination cites—and sometimes, within them—there is something more sinister rising: white nationalism and neo-Nazism.

The Ku Klux Klan, which people associate with the South, has long been in decline, while white nationalism, which is more associated with the North and West, is ascendant. The association of the Klan exclusively with the South is in fact a myth. While the Klan drew historically and spiritually from the South, white people across America flocked to the group, particularly at its height, including in destination cities and states.

That said, according to the Southern Poverty Law Center, the Klan has shrunk to a diminutive five thousand to six thousand members, a far cry from the millions of members it boasted at its height in the 1920s and 1930s. In 2019, there were three times as many white nationalist groups in America than Klan groups. These white nationalist groups are sprinkled all over the country, and not always where you might expect: there are none in Mississippi, while California boasts more than any state in the country.

Similarly surprising was a remarkable study published in 2015 in the journal *PLOS ONE* that used Google search data, which the researchers wrote "are unlikely to suffer from major social censoring," to determine the most racist parts of the country. They found that "the most concentrated cluster of racist searches happened not in the South, but rather along the spine of the Appalachians running from Georgia all the way up to New York and southern Vermont."

And many of the members of these white nationalist groups are young, college-educated millennial men, many like the columns of young men with tiki torches marching through Charlottesville in their Unite the Right rally in 2017 chanting, "The Jew will not replace us."

The rally was organized by then-thirty-four-year-old Jason Kessler, a local with a bachelor's degree in psychology from the University of Virginia, and then-thirty-nine-year-old Richard Spencer, who lives in Whitefish, Montana, has a bachelor's in psychology from the University of Virginia, a master's in the humanities from the University of Chicago, and was a doctoral candidate at Duke. The man who killed Heather Heyer during that rally, James Alex Fields Jr., was from Ohio; described by his high school teacher as "misguided and disillusioned," he had once written a paper that "was very much along the party lines of the neo-Nazi movement." Fields joined the army after high school but was released from active duty just four months later for "failure to meet training standards."

Kessler was a former member of one of the newest, headline-grabbing groups: the Proud Boys, something of a violent gang that describes itself as "Western chauvinist," and whose leaders "regularly spout white nationalist memes," according to the Southern Poverty Law Center. The group was founded by Gavin McInnes, whom the *New York Times Magazine* described as having gone "from Brooklyn Hipster to Far-Right Provocateur." McInnes is the cofounder of Vice

Media and *Vice* magazine and studied at Carleton University in his Ottawa hometown, then at Concordia in Montreal.

Today you could say that white supremacy is part of the higher education experience. In 2015 four out of five reported on-campus hate crimes were motivated by race, religion, or sexual orientation, with race motivating the larger percent of the three. And, according to the Anti-Defamation League, white supremacists targeting college students with leaflets ramped up in 2016, and by 2019, one-fourth of the total 2,711 white supremacist leaflet distribution events in that year took place on campuses.

Part of what's radicalizing young white men in destination cities and states, it seems to me, is a backlash against a decade of being chastised into checking their privilege and feigning shame or contrition over their historically oppressive identity. This is a backlash against a backlash, privilege rearing to battle antiprivilege.

New York City and other destination cities have in effect engaged in their own backlash, this one in the form of public policy ethnic cleansing—a sort of eradication by economic bullying, a removal of less desirable citizens for more desirable ones.

Washington, DC, was sued in 2018 for a billion dollars over gentrification in what the complaint called a twelve-year policy to "attract the Creative Class." The complaint charged that the city's "agenda disparately impacts other protected classes such as race, family, religion and matriculation." It went on to detail that some Black residents "are highly offended by current plans that are underway to actively racially integrate their neighborhoods, as if something is wrong with African-Americans living together in tight knit communities."

In 2015, Washington, DC, lost its designation as a majority-Black city, after being majority Black for decades—in 1970, the city was more than 70 percent Black. As one Black resident told the *Washington Post*, "D.C. was called Chocolate City . . . now, it's Chocolate Chip City."

Cities like the swap—younger, wealthier, educated white people likely to use fewer services and add more to the tax base in place of poorer people who use more services and earn less—so they help facilitate by aiming more police pressure at young Black and brown men so that the young white gentrifiers feel safe. This may not be an articulated plan but this is precisely how it functions: a pattern—dare I say program—of replacement through absorption and expulsion, in which young white people are wooed into gentrification and young Black ones are terrorized into evacuation.

In 2011, the Census Bureau released local data for New York that showed the first drop in the Black population of New York City on a census since at least 1880. (As a point of reference, the Reconstruction era ended in 1877.) The city's white, Asian, and Hispanic populations are all growing.

There are multiple forces pushing Black people out of destination cities, including a hostile criminal justice system, gentrification, and rising cost of living. But I contend that they are all related.

Perhaps no tool in the modern white terror playbook has been more effective, or notorious, than stop-and-frisk—a po-

THE PUSH

lice practice in which police are supposed to reasonably suspect that a person is armed before stopping him or her. The policy, an outgrowth of "Broken Windows" policing and the militarization of local police forces, spread like a plague across the country to include cities like New York, Los Angeles, Philadelphia, and Chicago.

The Columbia Law School professor Jeffrey Fagan produced a report that became part of a class-action lawsuit against New York City in 2010. It found that "seizures of weapons or contraband are extremely rare. Overall, guns are seized in less than 1 percent of all stops: 0.15 percent. . . . Contraband, which may include weapons but also includes drugs or stolen property, is seized in 1.75 percent of all stops."

As Fagan wrote, "The N.Y.P.D. stop-and-frisk tactics produce rates of seizures of guns or other contraband that are no greater than would be produced simply by chance."

When stop-and-frisk hit its peak in 2011, there were 685,724 stops, according to the ACLU. As the group explained:

Young black and Latino men were the targets of a hugely disproportionate number of stops. Though they account for only 4.7 percent of the city's population, black and Latino males between the ages of 14 and 24 accounted for 41.6 percent of stops in 2011. The number of stops of young black men exceeded the entire city population of young black men (168,126 as compared to 158,406). Ninety percent of young black and Latino men stopped were innocent.

That meant that if my boys, who were fourteen and seventeen at the time, weren't being stopped, other Black boys across town were taking those stops for them. In 2012 the *New York Times*' Op-Docs featured a young man named Tyquan who had been stopped between the ages of fifteen and eighteen "60 or 70 times" or "four or five times in a month."

A 2010 investigation by the *New York Times* found that "some eight odd blocks" in the Brownsville section of Brooklyn, just four miles from the tony neighborhood of Park Slope, "between January 2006 and March 2010, the police made nearly 52,000 stops," which amounted to "nearly one stop a year for every one of the 14,000 residents of these blocks."

Make no mistake: these young men—in many cases, just boys—were being hunted. We all felt like we were being hunted. As then-mayor Michael Bloomberg, a man who would seek the Democratic Party's presidential nomination, would put it in a speech at the Aspen Institute in 2015: "Ninety-five percent of your murders—murderers and murder victims—fit one M.O. You can just take the description, Xerox it and pass it out to all the cops. They are male, minorities, 16 to 25. That's true in New York. That's true in virtually every city."

The indiscriminate nature of the targeting of Black and brown men was part of the reason that my shoulders would draw up and the back of my neck would tighten and tingle when police officers were near. I tried to control it, but all efforts proved futile. This was an involuntary response to an ambient terror.

And yet, at the height of stop-and-frisk in New York City, a Quinnipiac University Poll asked residents twice in 2012 whether they agreed with the practice and both times a majority of white people said that they did.

The next year Quinnipiac asked New Yorkers if they thought

stop-and-frisk was "excessive and innocent people are being harassed, or do you think stop and frisk is an acceptable way to make New York City safer?" The majority of white people found it acceptable, twice as many as found it excessive. It wasn't happening to them or their sons. They had no skin in the game.

Bloomberg, the epitome of the white moderate, was so convinced of his virtue and certitude that Black pain didn't register in his massive expansion of the program. He vigorously defended the practice, even complaining in 2013 that "I think we disproportionately stop whites too much and minorities too little." As *USA Today* pointed out at the time: "About 5 million stops have been made during the past decade. Eighty-seven percent of those stopped in the last two years were Black or Hispanic."

I have never felt police anxiety in the South, in Atlanta or New Orleans or Jackson. Even when the officer stopped me and my friend in Ruston, I had seen him as an outlaw from the system rather than a representative of it. In those places, the cultures are simply different. They have majority-Black police forces.

Indeed, many of the largest cities in the South now have, or have recently had, Black chiefs of police: New Orleans, Baton Rouge, Shreveport, Jackson, Montgomery, Birmingham, Macon, Savannah, Athens, Columbia, North Charleston, Charlotte, Raleigh, Norfolk, Richmond, Baltimore.

Not all those cities have majority-Black police forces, but many have achieved an extraordinary level of diversity. Conversely, the percentage of Black officers in police forces in many destination cities is often shockingly low, and the percentage of white officers is out of sync with their percentage of the population.

In 2013, only 32.9 percent of the population of New York City was white, but 52.2 percent of the police force was. In Chicago, 32 percent of the population was white, but 52.1 percent of the police force was. In Philadelphia, 36.3 percent of the population was white, but 56.8 percent of the police force was. In many of these cities there exists a profound disconnect between those doing the policing and those being policed, each being somewhat foreign to the other, and in that gap grows fear, mistrust, and mercilessness. This can have disastrous consequences. Researchers at Texas A&M University analyzed randomly dispatched 911 calls and found that while white and Black officers "use gun force at similar rates in white and racially mixed neighborhoods, white officers are five times as likely to use gun force in predominantly Black neighborhoods."

This is not to say that the presence of Black officers is a panacea. It is not. The New Orleans Police Department is nearly 60 percent Black and that department has been plagued by bad behavior. Policing as an entity is a creature of culture and is somewhat congenitally hostile to Black people in this country, regardless of who wears the uniform. In many areas it is born of and shaped by the slave patrols and night watches, all designed to contain and control Black bodies, an inveterate instinct hard to breed out of the beast.

The officer who detained my son at Yale was Black. So were the police chief and the dean of Yale College, the first Black person to hold the position.

Brad W. Smith, an assistant professor of criminal justice at Wayne State University, has looked into the impact of police department diversity on officer-involved shootings and found that "measures of minority representation had no significant influence on levels of police violence."

But he found a difference between large cities (those with more than 250,000 residents) and smaller cities (those with 100,000 or more residents). In smaller cities, the actual prevalence of violence made more of a difference. In larger cities, the sheer number of Blacks made the bigger difference, "whereas community violence becomes nonsignificant."

As Smith put it:

Blacks in urban America are highly segregated and impoverished. These large, "threatening" populations concentrated in the nation's largest cities may produce much higher levels of antagonism between disadvantaged groups and the police. This may result in police perceptions of a general threat from citizens of impoverished minority communities. In smaller cities and large suburbs, on the other hand, ghetto communities are not as extensive or insular, and minorities may not be seen as a general threat. Thus a more generalized perception of threat may be less pervasive; rather, officers may respond to more specific threats from crime.

Put another way: Black density in these cities invites police violence and death.

The largest cities in America are more likely to be destination cities. Of the twenty largest cities in America, only six are in the South. Five of those are in Texas—Houston, San Antonio, Dallas, Austin, and Fort Worth—and the sixth is Washington, DC. All of those were internal destination cities.

Furthermore, the leadership of police unions is often white, even if the majority of the force isn't. A 2020 analysis by the Marshall Project found that "of the 15 large American cities in which a majority of officers are non-white, only one, Memphis, has a union leader who is Black."

In a way, police officers simply see the issue of race differently from most Americans. A 2017 Pew Research Center report found nearly all white officers, 92 percent, believed that "our country has made the changes needed to give Blacks equal rights with whites." Only 57 percent of the public overall believes that. Black officers were far less likely to agree with their white counterparts, with only 29 percent believing the country has made the necessary changes, but even this would be out of balance with Black people overall, among whom only 12 percent believe that the necessary changes have been made.

And the differential in the way police officers treat citizens and relate to them has a particular anti-Black tinge. The survey also found that while 91 percent of officers say police have an excellent or good relationship with white people in their communities, only 56 percent say the same about police relations with Black people.

Oppressive policing isn't only about mass incarceration and police shootings. Many local municipalities experience budgetary pressure. Rather than raise taxes or cut services, which are often politically unpalatable, they turn to law enforcement and the courts to make up the difference in tickets and fines. Some also increase the number of finable offenses and raise the fines and stiffen the penalties.

These fines and penalties are enormous sources of revenue for a city. In 2016, New York City raised nearly a billion dollars from fines, and that was up 16 percent from just four years earlier, even though the Census Bureau estimates that the city's population grew only 5 percent over that time.

Officers, already disproportionately deployed and arrayed in so-called high-crime neighborhoods—invariably poor and minority neighborhoods—are then charged with doing the deed. The increase in sheer numbers of interactions creates friction with targeted populations and ups the odds that individual biases will be introduced and inflamed.

Without fail, something eventually goes horribly wrong.

And then we train our focus on the interaction, examining the officers for bias and the suspect for threatening behavior, rather than looking at the systems that led to the interaction.

Society itself is to blame. There is blood on everyone's hands, including the hands still clutching the tax revenue that those cities needed but refused to solicit, instead shifting the mission of entire police departments "from 'protect and serve' to 'punish and profit,'" as *Mother Jones* magazine put it in a fascinating 2015 article on this subject. The magazine dubbed this police revenue chasing "policiteering."

Is it a coincidence that many of the cases involving Black people killed by the police began with stops for minor offenses? George Floyd was arrested and killed after allegedly using a counterfeit $20 bill.

This "fiscal menace," as the magazine called it, exerts pressure on a system often already addicted to ever-improving crime numbers—a statistically unsustainable condition—and a ballooning for-profit prison population, which the practice helps feed.

During Bloomberg's mayoralty in New York, tens of thousands of people were arrested for marijuana possession each year as a result of stop-and-frisk. There were more marijuana arrests under Bloomberg than under his three predecessors combined, according to an analysis by Harry G. Levine, a sociology professor at Queens College. Half of those arrested were Black and about a third were Hispanic. About 10 percent were white, even though research consistently finds that Blacks and whites use marijuana at about the same rates. If convicted, these Black and brown people were left with permanent criminal records.

The apparatus was rigged. As Levine put it in a 2012 Drug Policy Alliance report, every year, the police department was "stopping and frisking more than a half million mostly young Black and Latino men and falsely charging them with marijuana possession in public view."

Bloomberg knew the damage he was doing, but rationalized it away. As he defended the practice in his speech at the Aspen Institute: "One of the unintended consequences is, people say, 'Oh my God, you are arresting kids for marijuana, they're all minorities.' Yes, that's true. Why? Because we put all the cops in minority neighborhoods. Yes, that's true. Why do we do it? Because that's where all the crime is."

Dragging people into the court system creates another revenue stream by exposing those accused to a raft of fees, many of them mandatory. In 2017, New York City saw nearly $100 million in court-imposed fees and fines by its criminal and supreme courts, two-thirds of which were for misdemeanors, violations, infractions, and summonses. In 2018 families and friends transferred $17.5 million to the accounts of those in New York City jails; $13 million of that was spent in the jails' commissaries.

There exists a whole economy to bleed cash out of Black flesh, to produce revenue out of Black suffering, and all of this further drains already struggling neighborhoods of wealth and opportunity.

And to maintain the momentum of fines and arrests, cities have cracked down on lower and lower-level crimes, sacrificing more and more lives—largely poor and minority ones—to the apparatus. Public safety gives cover for a perversion of justice.

This fiscal menace isn't confined to the people ensnared; it is passed on to their families. According to the Prison Policy Initiative, a nonprofit, nonpartisan research and advocacy group, 70 percent of the people held in jail have not been convicted of a crime. On any given day, this is nearly half a million people.

America is essentially operating a network of debtors' prisons. In 2017 in New York, $268 million in bail bonds was posted and another $53 million was posted in cash bail. For people using commercial bail companies, they paid $27 million in nonrefundable fees, and the city reaped $15 million in forfeited or abandoned cash bail.

These bails are often paid by minority women. A 2018 report by the UCLA's Ralph J. Bunche Center for African American Studies found that the Los Angeles Police Department "levied $19,386,418,544 in money bail on persons arrested by the LAPD between 2012 and 2016." As the report pointed out,

"The estimated $19.3 billion paid in nonrefundable bail bond deposits were disproportionately paid by women, namely Black women and Latinas. Moreover, each community likely paid much more when accounting for post-arraignment payments, the service fees charged by bail bond companies, and, in some cases, asset seizures."

And, most of those who can't afford bail, who remain behind bars, are parents. This is a form of family separation that predates and vastly outnumbers the atrocity that has played out at the country's southern border with immigrant families.

Municipal officials evince no concern for the damage they're inflicting. Indeed, damaging the community produces more people likely to be ensnared by the system. They create and maintain the cycle from which they profit.

In the discussion of mass incarceration, people often point to the extraordinarily high incarceration rates of southern states. Eight of the ten states with the highest imprisonment rates, after all, are Louisiana, Oklahoma, Mississippi, Arkansas, Texas, Kentucky, Georgia, and Alabama, in that order. The other two are Arizona and Missouri.

Take my home state of Louisiana, for example. As the New Orleans *Times-Picayune* exposed in an extraordinary 2012 eight-part series:

> *Louisiana is the world's prison capital. The state imprisons more of its people, per head, than any of its U.S. counterparts. First among Americans means first in the world. Louisiana's*

incarceration rate is nearly triple Iran's, seven times China's and 10 times Germany's.

In the early 1990s, the state was under a federal court order to reduce overcrowding, but instead of releasing prisoners or loosening sentencing guidelines, the state incentivized the building of private prisons. But, in what the newspaper called "a uniquely Louisiana twist," most of the prison entrepreneurs also happened to be rural sheriffs.

That's right: the sheriffs own for-profit prisons. Talk about a perverse conflict of interest.

Beyond imprisonment, in Caddo Parish, the parish in which I was born, "from 2010 to 2014, more people were sentenced to death per capita [there] than in any other county in the United States, among counties with four or more death sentences in that time period," the *New York Times* revealed in 2015.

And yet in 2016, the Sentencing Project looked at the data differently, separating out incarceration rates in each state by the race of the people incarcerated. When you look exclusively at African-American incarceration rates per capita in state prisons, only two southern states, Oklahoma and Texas, are in the top ten. The other eight are in the North, West, and Midwest. Most of those states are among the whitest states in the union. When you look at Black men alone, the picture is even more stark. Vermont is the leader, incarcerating one in every fifteen Black men in the state.

Vermont is an interesting case in this regard: it reinforces the fact that white liberalism isn't always aligned with racial egalitarianism.

One of the most liberal states in the union, it also has one of the smallest percentages of Black people. Barack Obama won a larger share of the vote in Vermont than in any other state outside of Obama's home state of Hawaii in both 2008 and 2012. But, a 2018 Vermont Public Radio poll found that most Vermonters, 53 percent, believed that racism was a problem in the state, and 40 percent, a plurality, thought that not enough was being done about it.

A month before that poll was taken, Kiah Morris, the only African-American woman in the Vermont House of Representatives, resigned her post due to racist harassment and threats. She told the *New York Times* that her home was vandalized and invaded, she found swastikas painted on trees near her house, and she and her husband were the targets of racist social media posts.

The ugliness of Vermont's duality has crept into my own home. From 2015 to 2019 my son Ian was a student at Middlebury College, where he was a member of the football team. This is the same school that in 2017 became the poster child for radical liberalism in higher education when students shouted down the white nationalist Charles Murray and chased him from a speaking event, chanting "Racist, sexist, anti-gay, Charles Murray go away!" Just a few months after the protests, following the lead of Colin Kaepernick, my son decided to kneel at one of his football games during the national anthem to protest racial injustice. On October 28, he did so again, this time at a home game against Trinity College, and a man began to heckle him, yelling "Nigger" from the bleachers. As Ian told the *Middlebury Campus* newspaper, "As soon as he said, 'nigger,' I was immediately sure of what I was doing."

This is an odd artifact for a state that was home to some of the most radical abolitionists during slavery. As the *Burlington Free Press* has reported:

> Abolitionists launched the American Anti-Slavery Society in 1833, and less than six months later, 120 Vermonters gathered in Middlebury to establish their own society. By 1837, Vermont counted more than 90 local anti-slavery societies boasting more than 10,000 members.

All of this underscores my contention that racial egalitarianism is often, by necessity, a long-distance love affair.

The data reveal that the Black incarceration rate in Pennsylvania and California is higher than in Louisiana; the Black incarceration rates in Michigan, Missouri, and Illinois are higher than in Alabama; and the Black incarceration rates in Connecticut and New Jersey are higher than in Georgia and Mississippi.

When you zero in on some of the destination cities, the picture gets even worse. Just 5 percent of the people in San Francisco are Black, but Black people account for 56 percent of the people in the city's prisons. As San Francisco public defender Chesa Boudin has pointed out: "The reality is that San Francisco incarcerates African-Americans at a rate of more than 1,700 per 100,000 residents. For context, that's three times the incarceration rate of Russians at the height of the Gulag. It's the highest incarceration rate of African-Americans per capita of any major city in the country."

The Great American Mythos holds that racism is sectoral rather than omnipresent, pervasive, and suffused. In truth, there is no promised land of egalitarian actualization, no regional respite from racism. Black people in America must engineer their own Zion. They must carve it out of the hide of the country.

And yet, in the decades preceding the Great Migration, not even racial oppression and racial terror were enough to dislodge Black people from the South. No matter how much they suffered, the vast majority still stayed. They needed a nudge, a shock to the economic system to jolt them out of migratory stasis and into action. That shock came with a boll weevil infestation in cotton states beginning around the turn of the twentieth century. The combined forces of an economic crisis and a social justice crisis finally set the migration into motion.

I believe that we are seeing a similar scenario with the economic crisis created for Black people by the COVID-19 pandemic. First, as of August 2020, the age-adjusted mortality rate from the disease for Black people was three and a half times as high as it was for white people, and of the ten states with the highest Black mortality rates per capita, only one was in the South: Louisiana, at number six. The others, in descending order, were New York, Michigan, New Jersey, Connecticut, Massachusetts, Illinois, Pennsylvania, Wisconsin, and Minnesota. The District of Columbia would have been seventh on the list if it were a state.

At the same time, in the second quarter of 2020 (reflecting the period when the country's economy had essentially shut down) the national unemployment rate hit a historic high, with the Black unemployment rate higher than any other racial or ethnic group's. And, the Black unemployment rate was highest in the destination states of Michigan, Ohio, Pennsylvania, and Illinois, respectively. California, New Jersey, and New York tied for the fifth slot. The Black unemployment rate in Michigan hit 35.5 percent, roughly three times the rate in Georgia, which had one of the lowest rates in the country. (Data were not available for all states.)

In a grand twist of irony, the staggeringly incompetent response to the crisis by Donald Trump, a white power president, may well provide the necessary accelerant for a Black power migration.

Black people moved north and west during the Great Migration as the result of a combination of a push and a pull. The push was the racial terror and racial oppression of the Jim Crow South. The pull was the hope of a better life: economic, social, and political. I say the same sense of terror and oppression that pushed people out of the South has been reincarnated in the North and West. Hypermilitaristic policing, predatory incarceration, and the rebirth of a hate group movement are rendering destination cities unwelcoming, inhospitable, and, in some cases, uninhabitable.

Among Black people in these destination cities, there had long been—and still exists—a fetishizing of radicalism and rage in literature, education, and activism. I not only understand this affinity for rage, I'm attracted to it and believe it wholly justified. As James Baldwin is reported to have said, "To be a Negro in this country and to be relatively conscious is to be in a rage almost all the time."

Rage has always felt more active and empowering in the

North than it did (and does) in the South. But as I noted earlier, rage also fosters a combustibility that is foreign to many southern-born Black people like me. I know well that rage is an expensive emotion: it requires an untenable amount of continuous energy for sustenance; it can consume the very heart that holds it.

And although rage has often been an effective tool to focus attention and shift narratives, it rarely produces policy gains or positively shifts societal perspective. The beneficiaries of Black rage are often the more moderate figures with whom those in a rage are compared. Martin Luther King owes part of his success to Malcolm X, whom many whites saw as a more dangerous, and less acceptable, alternative.

As mentioned earlier, in the 1960s this Black rage resulted in riots in many northern and western destination cities, often over dire living conditions and oppressive policing. Fifty-odd years later, protests have once again broken out in many of the same cities, for the same reasons.

Instead of pursuing truly revolutionary change, too much Black energy, both activist and intellectual, has been too obsessed with finding ever-newer phrasings to articulate an old phenomenon and often doing so in service of explaining to white liberals what Black people already know as a matter of lived experience.

And I fear that Black activism is creeping toward its own form of elitism, a way of building strata and hierarchy of the supposedly woke over the supposedly asleep. Too many of our most lauded thinkers, most in the North and West, have rendered beautiful mediations and delivered blistering orations on the subject of Black liberation. But in the end, many succumb to a certain monotony of urbanity and arrogance, a plaintive howling into the wind, the building of a case without action, the diagramming of a problem without a solution. Much of it amounts to sullenness wrapped in sophistry.

Activism becomes an exercise in credentialing, a way of positioning in pursuit of power. These missives often represent as desperate longings by the authors to be anointed by white liberals and the white academy, collectives that address Blackness from a clinical distance, turning Black struggle into anthropology and Black pain into pedagogy.

It seems to me that on the racial question the white liberal has a nearly insatiable hunger for guilt-laden self-flagellation. In the same way, the white conservative has a thirst for absolution from legacy guilt and affirmation of current contempt.

The markets to appease both are robust.

But Black colonization of the South isn't a philosophy or an intellectual posture. It's an actual plan.

FOUR

THE PULL

THE POWER OF DENSITY

● ● ●

THE POWER OF THE BALLOT WE NEED IN
SHEER SELF-DEFENCE,—ELSE WHAT SHALL
SAVE US FROM A SECOND SLAVERY?

—W. E. B. Du Bois, *The Souls of Black Folk*

A hulking frame in a black T-shirt sits among four other men, two to each side of him behind two long tables, pushed together, and covered with tablecloths. The man in the middle is the rapper and activist Killer Mike, a prominent surrogate for Bernie Sanders, who has just lost a bruising nomination battle with Hillary Clinton. The panel has been assembled for the All Black National Convention in Atlanta, Georgia, in September 2016.

Killer Mike takes the microphone and begins to question the crowd about "revolutionary-ism." "Who fishes? Who knows how to fish?" he asks. Some hands go up. "Who knows how to hunt?" he asks. Fewer hands go up. This cycle repeats itself after every question. "Who shoots on a regular basis, meaning once or twice a week?" "Who farms and grows their own food, right now?" His voice grows more severe as he admonishes: "You ain't ready to oppose nothing. You are as a part of this system as any white person gentrifying in this city." And a little later: "I love you enough to tell you, you ain't ready to revolt shit." As his voice rises and the crowd becomes more excited, he leans into the microphone: "So, if you're not going to revolt tomorrow, if you're not going to do like Elijah said and take all the southern states, if you're not going to gentrify

Alabama, stop trying to grow warriors to fight a fight you're too scared to fight."

Killer Mike's message, beyond his self-sufficiency and survivalist impulses, resonates for me. I believe that revolution must extend beyond household self-sufficiency and pure economics and reach high into the power structures that govern. I believe that central to Black liberation is the assumption of power, the constitutional seizure of it.

I believe this must be central to the pull of Black people returning south.

The primary pull of the Great Migration was economic opportunity, as it was for the California Gold Rush during the mid-1800s and the Dust Bowl migration during the Great Depression. Surely, civil liberties and cultural tolerance also played a role in the decisions of Black people to migrate, but a first concern for any person is a roof over one's head and food on the table. They want and need to make the best living possible for self and family.

Similarly, a primary pull of the reverse migration is likely to be economics, particularly as many destination cities become increasingly unaffordable. The South provides this beacon of possibility as well.

In 2018 *USA Today* published an analysis by *24/7 Wall St.* of the "15 worst cities for black Americans," as measured by the social and economic disparities between white and Black residents across a variety of measures—income, poverty, adult high school and bachelor's degree attainment, home ownership, unemployment rates, incarceration rates, and mortality rates. Every city on the list was a destination city in the North and West—Waterloo, Iowa; Milwaukee, Wisconsin; Racine,

THE PULL

Wisconsin; Minneapolis–Saint Paul, Minnesota; Peoria, Illinois; Elmira, New York; Decatur, Illinois; Niles–Benton Harbor, Michigan; Kankakee, Illinois; Fresno, California; Springfield, Illinois, Trenton, New Jersey; Danville, Illinois; Rochester, New York; and, Chicago, Illinois.

Conversely, the South now looks like the land of opportunity that Blacks once expected to find in the North. In 2018 *Forbes* compiled a list of "The Cities Where African-Americans Are Doing the Best Economically," and found many of those cities were in the South, where rates of home ownership and self-employment were high and where Black household incomes were relatively strong.

When the Reason Foundation, a libertarian think tank, released its index of large American cities in 2019, southern cities topped the list as the most free, and northern and western destination cities were at the bottom as the least. The foundation's "Economic Freedom" index is primarily a measure of regulation and whether "a largely unregulated system leaves individuals maximally free to pursue their own plans, spurring entrepreneurial activity and innovation."

The "most free" metropolitan areas included Houston, Jacksonville, Tampa–Saint Petersburg, Richmond, Dallas–Fort Worth, Nashville, Miami, Austin, Orlando, and San Antonio. The least free metropolitan areas were Riverside (California), Rochester, Buffalo, the New York City area, Cleveland, Columbus, Portland, Sacramento, Providence, and Los Angeles.

A study by *Blacktech Week* of the best places for Black-owned businesses in 2017 found that "Southeastern states have a higher concentration of Black businesses." Of the top ten metropolitan areas with the highest concentration of black businesses, Georgia and Virginia each contained two, and Tennessee, Alabama, Louisiana, North Carolina, Maryland, and Florida had one each.

The top three metropolitan areas for Black-owned companies were Memphis, Montgomery, and Atlanta.

A 2019 Brookings report looked at the rise of Black household incomes across the United States and found that the largest statistically significant increases were in the West and South, although many of the western metro areas had small Black populations. The strongest growth was in the San Francisco Bay Area, where Black people represent just 5 percent of the population.

Among the metro areas with the largest Black populations—New York, Atlanta, Chicago, Dallas, and Washington, DC—the strongest growth was in Atlanta, nearly 70 percent more growth than Chicago.

And, the economic benefits are not only for the highly educated and well off. A 2020 study by economics professor David Autor at MIT looked at two big cities in each of the four quadrants of the country—New York and Philadelphia in the Northeast, Chicago and Detroit in the Midwest, San Francisco and Denver in the West, and Houston and Atlanta in the South—comparing real wage levels to city-specific price indexes from 1980 to 2015. He found that "accounting for regional price levels, the real wages of non-college workers fall in six of the eight cities in this period. Only in the southern cities of Houston and Atlanta do non-college wages make any net progress in these three-and-a-half decades."

More precisely, Autor observed that "Black workers in the South are not faring nearly as poorly as popular stereotypes might suggest. Among Blacks without a college degree, wage growth in the South is somewhat stronger than in the Northeast or West, and much stronger than in the Midwest. Among Blacks with a college degree, wage growth is stronger in the West and Northeast than in the South, but the gap is not large, and again the Midwest compares unfavorably."

Beyond economic motivation, another part of the pull must be toward self-governance, to the creation of an interstate safe space in which Black people are not forced to petition for the acknowledgment of their racial equality but can simply claim it.

So much of the power in this country is assigned to and controlled by the states. Slavery was allowed and maintained on a state level. Jim Crow was established and maintained by the states. As the Tenth Amendment sets forth: "The powers not delegated to the United States by the Constitution, nor prohibited by it to the States, are reserved to the States respectively, or to the people."

Black America must ask itself: How is the lack of any state control tolerable? What could we accomplish with state control?

This idea of Black people relocating to alter the political landscape is something that white America has long feared. During the first wave of the Great Migrations, the federal government became so panicked about the possibility of "vast vote colonization plots declared to have been hatched by certain political leaders" that the Justice Department went so far as to direct the United States district attorney in Louisville to "ascertain the names, point of origin and destination of negroes or any other voters from Southern States, particularly Alabama and Kentucky, who recently may have passed through Louisville to Northern States."

This fear is well founded among those who root for Black failure to support their mythologies of Black pathologies. They will no doubt point to struggling, majority-Black destination cities like Chicago and Detroit as evidence that Black control bodes poorly on an administrative level while conveniently omitting the fact that it was white flight and white disinvestment that crippled those cities. And they will conveniently omit the Black administrative success stories like Atlanta.

Atlanta became a majority-Black city in 1970. In 1973, Maynard Jackson was elected the city's first Black mayor. He completed an expansion of the city's airport into a world-class facility, ahead of time and on budget, an airport that is now the busiest in the world, and he brought the Olympics to the city. Throughout he forced Atlanta and anyone desiring to do business with it to be more racially inclusive. He created the blueprint for the most effective use of affirmative action, producing massive Black wealth in the city and creating scores of Black millionaires. The city has had Black mayors ever since.

Jackson's strategy was straightforward: "I believe that politics is the last viable means of nonviolent social change for Black people in this country."

It is precisely for this reason that I am putting forth this proposal, although I am far from the first to do so.

As Carol Moseley Braun, only the second popularly elected African-American senator in US history, and the first African-

American woman, told me in Chicago, "The idea that we could take over a state and actually develop political power as a result of demographics is not a new one. It's been talked about for one hundred years, at least." She continued, "It's not such a bad idea, frankly, because, you know, you got to have the ability to say, 'We control this.'"

One year after the Detroit riots of 1967—an event many Blacks called a rebellion—and in the midst of the nascent Black power politics in the city, the Malcolm X Society convened a meeting of hundreds of Black radicals from around the country, and formed the Provisional Government of the Republic of New Afrika (RNA). The group drew some high-profile supporters, including Malcolm X's widow, Betty Shabazz (who would be elected one of the group's vice presidents) and the acclaimed writer Amiri Baraka.

The group called not only for reparations, but also for the creation of an independent country in "the subjugated land" of Louisiana, Mississippi, Alabama, Georgia, and South Carolina, states with large Black populations and which had been the sites of extreme oppression. Among the RNA's Declaration of Independence aims were the following:

- To free black people in the United States from oppression;
- To promote industriousness, responsibility, scholarship, and service;
- To build a Black independent nation where no sect or religious creed subverts or impedes the building of New Society, the New State Government, or achievement of the Aims of the Revolution as set forth in this Declaration.

The group planned to achieve these goals through the establishment of a separate, autonomous nation within a nation. My proposition differs.

I am not advocating for a Black nationalism, but a Black regionalism—not to be apart from America but stronger within it, through consolidation and concentration. The goal is not sedition but liberty.

The idea of creating a separate state for Black people, as it happens, far predates all of these discussions and was dreamed up by white liberals and debated from the American Revolution through the Civil War, as University of Cambridge lecturer Nicholas Guyatt pointed out in his elegant book *Bind Us Apart: How Enlightened Americans Invented Racial Segregation*. Guyatt highlights a 1795 speech by a Dartmouth College lecturer, Moses Fiske, which advocated that "the solution was to create 'a large province of Black freemen' in the West, in which former slaves would be 'prepared for citizenship' and 'formed into a state.' Separation would 'bring them to an equal standing in point of privileges with whites.'"

And yet, among other things, the RNA's timing was off. Its message was discordant with the mood and movements of Black people in America. The group was founded just as the Great Migration was drawing to a close. Decades of Black momentum was aimed away from the South, not back to it.

Today the reverse migration is already underway. Reporters have been chronicling the phenomenon for at least a decade; almost 82,000 Black millennials migrated south in 2014 alone. At that rate, more millennials alone will have moved back to the South in twenty years than moved north in the quarter century that spanned the first wave of the Great Migration.

THE PULL

Already, the reverse flow of Black migrants from the North into Georgia has helped pave the way for Stacey Abrams to win the Democratic gubernatorial primary, making her the state's first Black nominee for governor and the first Black woman to be a major party's nominee for governor in the country. She came within a whisper of actually winning the race, which would have made her the first African-American governor in the Deep South since Reconstruction, when Louisiana's lieutenant governor P. B. S. Pinchback was sworn in as governor of Louisiana for six weeks as the sitting governor stepped aside for his impeachment trial. Pinchback was the first Black governor America ever had, and it would be nearly 120 years before the country would have another.

The bulk of Georgia's population growth from 2010 to 2018 was driven by the core metro Atlanta counties of Fulton, DeKalb, Cobb, and Gwinnett. Since 2010, the white population in those counties increased nearly 4 percent. Among Black people it grew 17 percent.

Still, this reverse migration is nowhere near as robust and energetic as its predecessor. There are four times as many Black people in America in 2020 as there were in 1910. To match the scale of the Great Migration, for the Black populations in southern cities to swell by exponential numbers as those populations in northern cities did, the reverse migration needs an adrenaline boost.

There are many reasons why the reverse migration is more trickle than flood. One reason is that the flow to this point has lacked concentrated encouragement and active recruitment. Not only is there no focused effort by southern industry to acquire northern Blacks for employment, there are few prominent voices lending the movement any sense of moral verve. The reverse migration doesn't have the equivalent of a Robert Abbott, and it should.

I want to help provide that boost. I want to grease the skid, to make the argument, to get people comfortable with Black migration and Black majorities.

I have known the nurturing, protective aura of majority-Black places.

Shreveport, Louisiana, the town in which I was born, the town in which I worked my first newspaper internship, the town in which I met my ex-wife and where I married her, is majority Black.

Kiblah, Arkansas, the flyspeck community where I spent the first years of my life with my grandmother and her fourth and final husband, is majority Black.

My hometown of Gibsland, Louisiana, was majority Black, as was my high school. The neighboring town of Arcadia, where my mother once worked in a chicken-processing plant that is directly across the street from where one of my brothers now works in a precision alloy plant, where I saw my first dentist and my first doctor and my first movie, is majority Black.

The tiny cluster of homes between Bryceland and the now-vanished town locals call Old Sparta, the town that was once the home of Marshall Harvey Twitchell, a white man from Vermont who had been a captain of Company H, 109th Colored Infantry during the Civil War and who would be named provost marshal and agent of the Freedmen's Bureau in the wake of it, that surrounded Shiloh Baptist Church, where I was baptized and my father's father was buried, was majority Black.

THE PULL

The college I would attend, Grambling State University, was also majority Black and situated in a majority-Black town. Detroit, the city in which I held my first postcollegiate job, where my older son was born, is majority Black.

In 1994 I moved to Brooklyn, which during the 1970s outpaced Harlem's Black population, "becoming the largest black community in the Northern hemisphere" as WNYC wrote in 2011. I first settled in then-majority-Black Prospect Heights, part of the district that sent Shirley Chisholm to Congress, the first Black woman elected to the body and the first to run for president.

These places were not aberrant to me, but typical. And, many of the people I came to know from the South also came from majority-Black towns and cities.

There are now more than 1,200 majority-Black cities and towns in the United States, and more than 1,000 of them are in the South. There are almost nine million Black people in those majority towns. More than six million of those are in the South. This means that nearly a quarter of all Black people in America already live in majority-Black towns, and the overwhelming preponderance of those people are in the South.

There are as many Black people living in Black-majority towns in the South as roughly the population of the West African country of Sierra Leone. (Ironically, Sierra Leone was a major source of enslaved Africans transported to what would become the United States and as such the country is the most frequent result for DNA tests from Black people in America.)

These southern Black towns, merely as a matter of scale, essentially already constitute a country within a country, a reconstitution of Black control, a new Africa in America.

In fact, the South is proving particularly attractive to new immigrants from Africa. A 2017 Pew Research Center report found that the total US foreign-born population from Africa has skyrocketed, rising from about 80,000 in 1970 to over 2 million in 2015, and a higher percentage of African immigrants to the United States settle in the South than in any other region of the country.

Most of these majority-Black southern municipalities are small towns, but among them are some larger cities like New Orleans, Baton Rouge, Memphis, Birmingham, Montgomery, Atlanta, Savannah, Richmond, Washington, and Baltimore. Indeed, until recently the Blackest city with a population over 100,000 people was Detroit. Now that city is Jackson, Mississippi.

Significantly, many of these cities are being led by young, innovative, even revolutionary Black mayors. Half of the cities listed above have mayors under fifty. Washington's Muriel Bowser, forty-two when elected, and Richmond's Levar Stoney, thirty-five when elected, are the youngest mayors in those cities' histories. Randall Woodfin, thirty-six when elected, is the youngest mayor in Birmingham in over one hundred years. This young leadership and fresh ideas are what the next wave of Black leadership should look like.

It is particularly thrilling, as a Black person in America, to arrive at the airport in Jackson, Mississippi, and hear the voice of thirty-seven-year-old mayor Chokwe Antar Lumumba welcome you to the Jackson–Medgar Wiley Evers International Airport. Lumumba takes his name from his father, who was also mayor of the city, and who was a member of the Republic of New Afrika, and moved to Mississippi as part of that group's effort at colonization.

The possession of real statewide political power in the South could radically alter the architecture of oppression in this country.

For one, we could better deal with the idea of reparations, which has reemerged as an exigent issue, with statements by presidential candidates and congressional hearings.

Reparations is a mountain with two peaks: one moral and one legal. In the first instance, the white majority must be convinced that it is morally right for America to atone for its sins of slavery and oppression. In the second, and even more difficult ascent, the Congress must pass and a president sign a bill authorizing it, in whatever form it may take.

That is, unfortunately, a near impossibility considering our current politics. But one way to make it more tenable is to send more senators to Washington for whom it is a priority. Barring that, gaining control of states could allow for reparation on a more local level rather than federal.

As the Pew Trust has pointed out, four states—New York, Vermont, California, and Texas—have introduced bills to study or enact reparations. Florida lawmakers introduced a bill to pay reparations for a specific catastrophe: the 1920 Ocoee massacre, in which a white mob on Election Day rioted to prevent Black people from voting; differing reports put the number of African-Americans killed as low as three and as high as sixty. Unfortunately, the reparations provision was dropped in negotiation, and what passed was a bill to recognize the massacre in school curriculums. The city council of Asheville, North Carolina, voted unanimously to approve a reparations resolution that would make investments in areas where Black residents face disparities.

Black density wouldn't prove beneficial only for political reasons. Black people also need to reunite to combine purchasing power, brainpower, and cultural power. Our dispersal has exposed us to exploitation, for which density could be curative.

Black people are heavy users of social media, particularly as an instrument for activism. A 2018 Pew Research Center survey found that Black people, more than others, value social media as a way to amplify lesser-covered stories, to get involved with issues, to find others who share their views, to give voice to underrepresented voices, and to achieve political goals. But, not only are there no Black CEOs at the major tech companies, there are only a handful of Black executives, and the percentage of minorities employed by most of them is abysmal. According to a 2018 analysis by Recode, Microsoft, Intel, Twitter, Facebook, Pinterest, and Google, Black people represented less than 4 percent of their total workforces. When *Adweek* published its 2018 Power List of "100 cutting-edge marketers, media CEOs, brand champions and tech titans," there wasn't a single Black person on it.

The majority of players in the National Football League (the most profitable sports league in America) and the National Basketball Association are Black, 65 percent and 75 percent respectively—however, as of October 2020, there isn't a single Black owner of an NFL team, and only one Black

THE PULL

owner of an NBA one, Michael Jordan. The NFL just, in 2020, saw its first Black president of a team, with the appointment of Jason Wright by the Washington Football Team. These leagues generate billions of dollars in revenue, and only a portion redounds to the athletes and almost none to the Black communities that nurtured them.

Furthermore, historically Black colleges and universities, which still produce nearly a quarter of all Black graduates with an undergraduate degree, receive little of the benefit from these athletes' talents. Instead that Black talent draws billions of dollars of revenue to overwhelmingly white schools and media groups. The NCAA, television networks, and Division I schools—almost all majority white—make billions of dollars off of these athletes.

As Jemele Hill points out in *The Atlantic*:

> Bringing elite athletic talent back to Black colleges would have potent downstream effects. It would boost HBCU revenues and endowments; stimulate the economy of the black communities in which many of these schools are embedded; amplify the power of black coaches, who are often excluded from prominent positions at predominantly white institutions; and bring the benefits of black labor back to black people.

And then there's hip-hop, a towering creation and expression of Black culture, one influencing fashion, music, and style around the globe. In 2018, hip-hop/R&B surpassed rock as the most consumed music genre, and yet, as hip-hop mogul Sean "Diddy" Combs told *Variety* magazine, "You have these record companies that are making so much money off our culture, our art form, but they're not investing or even believing in us." Combs continued: "For all the billions of dollars that these black executives have been able to make them, [there's still hesitation] to put them in the top-level positions. They'll go and they'll recruit cats from overseas. It makes sense to give [executives of color] a chance and embrace the evolution, instead of it being that we can only make it to president, senior VP.... There's no black CEO of a major record company."

Black people have $1.2 trillion in purchasing power and Black talent produces untold billions for other people. How powerful would the Black community be if that money and talent remained and revolved internally, between Black people and Black business?

There is density in Black America now, to be sure, but it is spread over several northeastern states, rendering its power less effective. There are 3.4 million Black people in New York, the second-most-populous state. And yet the only majority-Black city in New York is Mount Vernon. (In California, the most populous state, there isn't a single majority-Black city.) The vast majority of New York's Black population, two million people, lives in New York City, but they make up only a quarter of the population there. In the hundred years since the Great Migration began, NYC has had only one Black mayor: David Dinkins, elected in 1989. The state has never elected a Black senator and had a Black governor for a few years only because the governor at the time was forced to resign after being caught in a prostitution scandal. There are three million Black people in New York State and the only real federal power they currently produce is three Black representatives from Brooklyn and Queens.

THE PULL

The city has had only two Black police commissioners, the last of whom left office in 1992. And yet, consider this: One out of every twenty Black people in America lives in New York City. One in every five lives in the Northeast Corridor stretching from the Washington, DC, metropolitan area to the New York City metropolitan area. That's only about one and a half times as long as the distance from the top of Alabama to the bottom.

Our current density is somewhat impotent. Black people don't just need absolute density, they need strategic density, centered in state and regional control.

Among the most prominent and powerful groups of Black people in America are those in Congress, but they have a generational problem. When the 115th Congress was sworn in in 2017, the average length of service for all members was 9.4 years. For Black members, it was 12.2 years. For Black members from the South, the number was 13.3 years. The average age of all members was 57.8 years old. Among Black members it was 63.3 years old. Among southern Black members, it was 66.7 years old.

The Black political power structure in this country, antiquated and sclerotic, is in desperate need of a jolt of youth and energy, and a wave of young migrants to the South who can establish state or regional control can provide it, in the spirit of the region's young Black mayors.

I am talking about a grand generational undertaking, a rescue mission for Black America.

And that mission begins with the states, which are the true

centers of power in this country, and as such control the lion's share of the issues that bedevil Black lives: criminal justice, judicial processes, education, health care, economic opportunity and assistance.

There is no issue more important to Black self-determination than criminal justice, from on-the-street interactions with police officers all the way to precedents set in the Supreme Court.

Controlling a state provides far greater power than any city could ever possess, so the great Black destination meccas like New York, Detroit, and Chicago are still beholden to states of which Blacks constitute less than 18 percent of the population.

There are no cities in the Constitution. Municipal power is conferred by the state and can be preempted by the same. Whatever laws a city enacts, including criminal code, can be overruled by the state, and state preemption of municipal power is on the rise. State control would allow for the alteration of the criminal code.

Much was made over the passage of the First Step Act in 2018, which, among other things, retroactively applied the Fair Sentencing Act to the incarcerated, making many eligible for release. But, the act applied only to federal prisoners. The vast majority of people incarcerated in America are in state prisons, 1.3 million in 2020. About half as many are in local jails, and only about a sixth as many are in federal prisons.

To put a meaningful dint into mass incarceration, state control is crucial. And there must be a greater role played by Black political power in prosecutions and convictions. In most cases not only are local district attorneys and state attorneys general elected officers, but so are local trial court judges, intermediate appellate court judges, and state supreme court justices.

There is also a persistent racial achievement gap in America, with research showing that white students score an average of 1.5 to 2 grade levels higher than Black children. This issue must be addressed on a state level because public education is primarily a state and local issue in America; more than eight out of every ten dollars of educational funding comes from the state and local levels. As the Center for Public Justice points out, "State governments exercise primary accountability and oversight for government-run schools," and "state-level funding for government schools varies widely, but on average, states provide for slightly less than half of schools' operating costs."

Black people aren't even educating our own children anymore. As *Education Week* reported in 2019, sixty-five years after *Brown v. Board of Education*, nationally only "about 7 percent of public teachers, and 11 percent of public school principals, are Black." Prior to the decision, 35 to 50 percent of the teaching force was Black in the seventeen states that had segregated school systems.

States also play a major role in health care, allocating about $300 billion of their direct spending—roughly equal to their spending on education—on the sector.

Black people are disproportionately affected by a spectrum of chronic diseases like heart disease, diabetes, hypertension, lung disease, and certain cancers. Robert J. Sampson and Alix S. Winter, both Harvard sociologists, wrote in a 2016 paper about the racial ecology of lead poisoning in Chicago, "The link between racial segregation and multiple social adversities is a central feature of the American landscape."

Nothing laid this fact bare more starkly than the COVID-19 pandemic, in which Black people in this country were more likely to die of the disease because of their preexisting conditions. And these effects were just as pronounced in destination cities as southern ones, if not more. As John C. Austin wrote for the Brookings Institution:

> *The pandemic is blowing away the illusion that racism in the North—manifested in practices such as redlining, deeded covenants and shifting public school boundaries when black children began to mingle with white children—was at least not as violent as the lynchings, fire hoses and fire bombings that characterized Southern racism. Almost overnight, the COVID-19 pandemic has turned historically institutionalized racism in the Midwest's industrial cities into a murder weapon.*

The impact of a state's decision, particularly on a health issue disproportionately affecting Black people, can be devastating, as we've seen with the HIV epidemic ravaging the South. As the US Centers for Disease Control reported in 2016:

> *In the decades since the first AIDS cases were reported in Los Angeles and New York City in 1981, the epicenter of the nation's HIV epidemic has shifted from urban centers along the coasts to the 16 states and District of Columbia that make up the South. The South now experiences the greatest burden of HIV infection, illness, and deaths of any U.S. region, and lags far behind in providing quality HIV prevention and care to its*

citizens. *Closing these gaps is essential to the health of people in the region and to our nation's long-term success in ending the epidemic.*

A 2019 CDC issue brief pointed out that Black people in the South accounted for 53 percent of new HIV diagnoses in the country in 2017. And nearly half of all deaths from AIDS in the United States were in the South.

One reason this disease is spreading so rapidly among Black people in the South is because many simply don't have access to care. And yet, many of the southern governors refused to expand Medicaid under the Affordable Care Act, even though the program is a primary source of HIV treatment for many. As a 2016 Henry J. Kaiser Family Foundation fact sheet put it, Medicaid "is the single largest source of coverage for people with HIV in the U.S. and the number has grown over time," estimated to cover more than 40 percent of people with HIV in care. The report continued, "One analysis found that if all states expanded their Medicaid programs, nearly 47,000 people with HIV could gain new Medicaid coverage."

In 2015 Democrat John Bel Edwards, fueled by high Black turnout, won the Louisiana governorship. On his second day in office he signed an executive order expanding Medicaid under the Affordable Care Act and "fulfilling a campaign promise that will expand health coverage to hundreds of thousands of people in one of the nation's poorest states," as the *New York Times* reported. That year Louisiana had the highest rate of new HIV diagnoses of any state in the country. Most of those new infections were of Black people.

Fast-forward to 2018, and new transmissions had dropped

by 12 percent over three years, a dramatic decline that state health officials attributed to the expansion of Medicaid. These are lives saved because a state is controlled by someone sympathetic to Black lives.

Born in 1970, in the middle of the Black power movement, I have heard progressive, enlightened, activist Black people speak of Black advancement in revolutionary, Black nationalist language, all my life—"The revolution is coming," "The revolution will not be televised"—but the nature of this revolution and the likelihood of its success were always ephemeral, even fanciful.

Armed revolution was doomed to swift and brutal failure. Political revolution, which never seemed to be the major thrust of these longings, still hinged on convincing a white majority to defer. And a culture revolution—which Black people have influenced or ignited for as long as they have been in this country, even before it became a country—produced narrowly distributed economic power and almost no political power.

And so, revolution became an idle phrase of Black resistance, hollowed out by lack of definition and direction. We moved to a place where the revolutionary-ism was overtaken by the even less well delineated "radicalism," a term now so ubiquitously applied and awarded without circumspection that it has nearly been stripped of substance.

I think there is now a generational remove from a time when those words still carried weight, one to which we must

THE PULL

return. We got a glimpse of that possibility in the massive protests that sprang up in the wake of George Floyd's killing.

But, it seems to me that the most revolutionary act of self-determination Black people could take is the mass resettlement of the South.

I'm not saying that all Black people should or would participate. Mass movements are largely for the young and unencumbered; that was the case with the Great Migration, and that was the case with the mass movement to Vermont. The allure is that it's a revolution without violence. It uses the law, the Constitution, the very mechanisms that have been employed in Black oppression, as the vessel for Black liberty. For decades white supremacists have championed states' rights as tool of racial oppression. Here that philosophy will be employed in the quest for racial liberation. White supremacy is to be consumed by its own fire.

When I was about to graduate from high school, my intention was to go away to college, anywhere other than the local school, Grambling State University, the historically Black college just twenty miles down the road, the one to which almost everyone from my school went, the one to which my mother and three of my brothers had gone.

I was a boy in need of flight. I needed to get away, run away, to cast off the persona I'd long performed, to find the truth of me, the whole of me, and dwell within it.

My intention was not everlasting abandonment of my home state, but a respite from it. Since learning about the raucous, tumultuous history of Louisiana governors, and meeting the outlandish current occupant of the office, Edwin Edwards, I had set becoming the first popularly elected Black governor of the state as my aspiration.

In my limited research of how I might best bring this to fruition, I surmised that I should double major in English and pre-law, go to law school, become an attorney, and use that career as a springboard into politics.

From what I read, a small liberal arts college would likely be the best fit for me, so that is where I narrowed my search. For reasons I cannot recall, I settled on the College of William and Mary in Williamsburg, Virginia, over a thousand miles north and east of Gibsland.

I figured I had a decent chance of being accepted. I was the valedictorian, captain of the basketball team, had solid test scores, was founder and editor of the school newspaper, and was president of every club I was in, and also president of my class since the fifth grade.

But, I had no idea how I would pay the tuition. There my wanderlust was cut down by my lack of imagination and resources. I didn't even apply. The dream died in the same manner it was born, in silence.

So, I trained my sights on more provincial ambitions. I applied to Louisiana State University. If I couldn't go north, I would go farther south. I was offered a full scholarship. I also won a scholarship from Louisiana Tech University, just three miles from Grambling, but I had no intention of attending. Too close.

LSU, I believed, was the premier school among my options, the better choice. So, I chose it. But, the recruiter from Grambling had other plans. Before I formally accepted any offers,

THE PULL

he summoned me for a meeting in his office at the center of campus, in a small brick building that looked like it surely had once been a home. He asked me about my options and awards, I told him about the two scholarship offers. He told me that Grambling would also give me a full scholarship. Then, he gave me a speech about something greater than money, facilities, and prestige. He said, "LSU doesn't *need* you! Louisiana Tech doesn't *need* you! Grambling *needs* you!"

I understood immediately what he was saying: not that Grambling needed me for some curve-bending academic accounting, but that Black people needed me, that Black culture needed me, that my presence among my Black peers would be one of giving and inspiring as much as learning and being inspired. He was right, so I chose to strive with my people, to have less but gain more, to find a way to rise together.

I have never forgotten those lessons, that sometimes the mission is greater than you, that sometimes the decision you make for you must include the consideration of the whole, that while my presence in white space may be additive, that presence in Black spaces could be curative, that sometimes the choices we make in life are more about community advance than personal prerogatives. He was telling me that at times altruism is the greatest ambition.

Now I'm saying to today's young, gifted, and Black: Your people *need* you. Will you consider making the choice that I once made—not just in selecting a school, but in selecting a place to live out your life? Will you take it as a mission?

Black people have it within their power to be their own saviors, to craft their own liberty, to author a new narrative.

FIVE

THE END OF HOPING AND WAITING

● ● ●

FOR YEARS NOW I HAVE HEARD THE WORD
"WAIT!" IT RINGS IN THE EAR OF EVERY NEGRO
WITH PIERCING FAMILIARITY. THIS "WAIT"
HAS ALMOST ALWAYS MEANT "NEVER."

—Martin Luther King Jr.,
"Letter from Birmingham Jail"

A few weeks before Christmas in 2018 I sat at a table in a darkened Midtown Manhattan ballroom, the guest of a friend at the Robert F. Kennedy Human Rights Ripple of Hope Award Gala. It was the fiftieth anniversary of his assassination. Barack Obama was receiving the Ripple of Hope Award that evening.

Obama bounded to the stage. There was enthusiastic applause. He opened his remarks with a joke: "I'm not sure if you've heard, but I've been on this hope kick for a while now." The crowd laughed. He flashed his toothy, boyish grin, rewarding his own wit, and, as is often the case, unable to refrain from laughing at his own jokes. It is charming. He is charming.

He soon settled into the substance of his talk, in part defending his adherence to hope itself. He seemed to be debating an apparition, a dissenting voice within himself, trying desperately to support that which no one in the room was challenging.

It was a remarkable speech, eloquent and adroit. He demonstrated once again why some have termed his wunderkind mastery of rhetoric "Ciceronian" after the ancient Roman politician. It is in the way he effortlessly matches the mood of a room, how he absorbs adulation in a way that makes the giver feel honored in the giving.

He explained that hope was "never a willful ignorance to the hardships and cruelties that so many suffer or the enormous challenges that we face in mounting progress in this imperfect world," but rather an "insistence that no matter how tough our circumstances there are better days ahead," that "despite all of our individual failings and all of our inadequacies, together we can overcome" and "a belief in goodness and human ingenuity, and maybe most of all our ability to connect with each other and see each other in ourselves." He warned, "It can be tempting to succumb to the cynicism, the belief that hope is a fool's game for suckers," but that "Bobby Kennedy's life reminds us to reject such cynicism."

This passage stuck with me, as I demurred. It seems to me that the opposite of hope isn't so much cynicism but realism. Surely, hope can be a potent weapon against despair for the powerless and oppressed, for the wishful and the prayerful. But we must always stipulate that hoping isn't an exercise *of* power but an exercise *in* faith.

An overreliance on hope can itself become a hindrance. It's a sedative, a calming potion. It can substitute wishing for working. It can whisper in dulcet tones that out there, somewhere, beyond the point where the vision falters, in that nether place, lies the prize for your patience.

In the clearest consideration, it is a pleading and an imploring that power will bow to righteousness. Hope, as a religious tool, may well be essential; but hope, as a political tool, is folly.

And yet hoping is rooted in the Black experience in this

country and pervades Black politics. Historically it has consumed the rhetoric of Black politicians and the rhetoric directed at Black voters. Andre C. Willis, an assistant professor of religious studies at Brown University, describes African-American hoping as a tradition that has been "crafted over centuries of despair and dehumanization." Hope was an essential commodity among the downhearted, often the only light visible from the darkest of places.

Hope, and the religiosity to which it is wed, have been drilled into the collective Black consciousness since enslavement. And in the South, where enslavement was most entrenched and most prevalent, Black religiosity remains a prevailing psychology.

Black people in the South also were (and remain) more religious and more conservative than their counterparts in the North. I asked the pollsters at Gallup to run an analysis of religiosity by race and region of the country. Black people were the most religious of all races, and Black people in the South were more religious than in any other region. For instance, 54 percent of Blacks in the South were "highly religious" compared with just 39 percent in the West and 41 percent in the Northeast.

Religion anchored Black people's existence in something greater than themselves. The same Bible that white people had used to justify the enslavement of Black people, Black people would use to illuminate the immorality and spiritual depravity of white people's racial intolerance, animus, and oppression. Southern Blacks turned the white man's scripture and the white man's Jesus against the white man's hatred. Indeed, American Christianity didn't sanctify Black people; Black people sanctified American Christianity. They brought to it a living model of persecution, a perfect suffering and a perfect sustenance.

But Black people are still too heavily indoctrinated into a philosophy that promises comfort, justice, and peace in an afterlife, particularly to those who don't have it in this one. Karl Marx was right when he called religion "the opium of the people." So long as people are shouting in church they are less likely to shout in the streets.

My view of religion is simple: Human beings needed a way to consider things they couldn't comprehend, so they created religion. It's a way of relieving anxiety and enforcing a morality. Our brains are magical, so they invented magic.

But this magic now works against Black people in many ways. And, it's highly correlated to prosperity or the lack thereof. The wealthier, higher educated are less religious and the poorer, less educated are more religious. For instance, Asian-Americans have the highest incomes and highest educational attainment and have the lowest level of religiosity. African-Americans have the lowest incomes and the second-lowest educational attainment, and have the highest religiosity.

The Ripple of Hope Award Obama was receiving that December night took its name from a speech Kennedy delivered at the University of Cape Town, South Africa, in 1966, after the Civil Rights and Voting Rights Acts had passed in America, but while South Africa was experiencing its worst years of apartheid. In that speech, Kennedy talked about the "painful slowness" at which rights and freedom were being extended to Black people in America. He went on to say:

THE END OF HOPING AND WAITING

Each time a man stands up for an ideal, or acts to improve the lot of others, or strikes out against injustice, he sends forth a tiny ripple of hope, and crossing each other from a million different centers of energy and daring those ripples build a current which can sweep down the mightiest walls of oppression and resistance.

When Obama spoke at the same school in 2013 he invoked that quote by Kennedy and implored the crowd:

Think about how many voices were raised against injustice over the years—in this country, in the United States, around the world. Think of how many times ordinary people pushed against those walls of oppression and resistance, and the violence and the indignities that they suffered; the quiet courage that they sustained. Think of how many ripples of hope it took to build a wave that would eventually come crashing down like a mighty stream.

In fact, the events Obama referenced were not hopes or ripples, but actions. During his presidency he helped popularize the quotation, which was born of a nineteenth-century Unitarian minister, Theodore Parker, and paraphrased by Martin Luther King Jr., "The arc of the moral universe is long, but it bends towards justice." I say that it doesn't simply bend as a consequence of natural progression; it must be bent, with great force and at great cost. And, I say that the time for hoping and waiting, as a political strategy among Black people, must end. The path to power and relief from racial oppression is before us. We need to take it.

And yet, northern white liberals—as well as the many polished and proper Black people whose prosperity is born of proximity to that whiteness—tell the great masses of Black people who toil and struggle, whose love and devotion to the liberal cause too often go unrequited by liberal action, to hold fast and have hope. They tell them that the time will soon come when this great racial conflict will pass peacefully into the annals of infamy.

But where, precisely, should Black people place that hope?

The experts tell us that the browning of America and the rise of interracial marriage will one day make race itself a moot point. There is much talk about the racial majority-minority tipping point of America and of individual states and cities, but such talk often glazes over the fact that the issues and interests of individual minority groups don't always align, and indeed can wildly diverge.

It also fails to acknowledge that soon after reaching the majority-minority marker, many states will also see the ascendance of one group to majority status, namely Hispanics. Within the next thirty years, Hispanics will become the majority population of Nevada, Arizona, New Mexico, and Texas, and while California won't be majority Hispanic, Hispanics will be the largest racial group in the state. This will create for Hispanics their own regional power center. Hawaii has always enjoyed a majority or plurality of Asians and Native Hawaiians/Pacific Islanders. Most other states will maintain their white majorities—or at least a white plurality—for decades to come, particularly in the northwestern and plains states.

Maybe that's fitting. Many of the states that will maintain their white majority are in an area that was known as Oregon

Country in the early 1800s and included the present states of Oregon, Washington, and Idaho, and parts of Montana and Wyoming. In 1844, fifteen years before the state of Oregon was admitted to the union, the provisional government in Oregon Country passed a law that mandated that Black people leave the territory. The "Peter Burnett Lash Law," named after the former slaver who led the region at the time, was enforced brutally, mandating the severe whipping of Blacks who refused to leave.

This law was rescinded but replaced by a series of other racial exclusion laws, the last of which was not repealed until 1926.

Those who tell Black people to wait and hope also don't acknowledge the anti-Black racism that exists both within racial minority groups and among them. Project Implicit, a virtual laboratory maintained by Harvard University, the University of Washington, and the University of Virginia, has administered hundreds of thousands of online tests designed to detect hidden racial biases. The tests find not only that three-quarters of whites have an implicit pro-white, anti-Black bias, but also that nearly as many Hispanics and Asians share that pro-white, anti-Black bias.

Furthermore, in 2016, after Donald Trump had lobbed years of racist birtherism attacks against President Barack Obama, after he had called Mexicans rapists and murderers and after repeated promises to build the wall, after he claimed that Islam hates America and called for a ban of all Muslims coming into the country, he still won 28 percent of the Hispanic vote and 27 percent of the Asian vote, according to exit polls. He won only 8 percent of the Black vote.

During the 2018 midterms, after Trump had referred to Haiti and African countries as shitholes, after he had drawn a moral equivalency between Nazis in Charlottesville and those who came to protest against them, after he had instituted a draconian form of family separation at the southern border which left many Hispanic children in cages and some dead, after he had demeaned football players who kneeled in protest of police violence against Black bodies, Hispanics still voted for Republican candidates who almost universally supported and shielded Trump at roughly the same level as they voted for Trump in 2016, according to exit polls: 29 percent of Hispanics and 23 percent of Asians voted for Republicans, somewhat in line with Trump's 2016 results. Just 9 percent of Blacks voted for the Republican candidates.

A Quinnipiac University Poll conducted in March 2019 found that 29 percent of Hispanic voters said they would definitely vote, or consider voting, for Trump in 2020. In July 2020, in the midst of racial justice protests that swept the nation, the number of Hispanics who said they would vote for Trump actually increased to 35 percent. On Election Day, 32 percent actually did vote for him, according to exit polls.

A sad truth throughout the world—and specifically in this country—is that whiteness, aesthetical if not cultural, for many has become aspirational. It is hard to find a society anywhere in the modern era that has some diversity of phenotypes in which the darker ones are not assigned to the lower caste. Those beliefs follow people from their home countries to this one. It seems to me that we idealize a societal construct for which history offers precious few precedents. Most of the world seems to operate as some variation of a pigmentocracy.

If I was fully convinced that the browning of America would lead to the full equality and inclusion of Black people

THE END OF HOPING AND WAITING

in this country I wouldn't have written this book. My fear is that browning could mean that white supremacy is simply replaced by a form of "lite" supremacy, in which fairer-skin people perpetuate a modified anti-Blackness rather than eliminating it. From my experience, colorism is just as potent and dangerous a force as racism. Egalitarianism is not an evolutionary given. It is by no means destined. Tribalism is an intrinsic evolutionary trait. It's a neurological preset. We must consciously and continuously condition ourselves out of it. Furthermore, intersectionality does not guarantee a perfect ligament. An ally on one issue can be armed against you on another.

Black people have dealt with the lite-to-dark dynamic intraracially for centuries, with "house slaves" juxtaposed with field ones, with paper bag tests and blue vein societies, with those who passed for white and those who couldn't or wouldn't. The prospect of dealing with it interracially, among other historical minorities in this country, does not entice.

I understand the appeal of racial coalitions, but there is a preciousness in the idea, a laudable allure, that doesn't always manifest in the aggregate. I too dream of a world in which people come together across immutable differences, including race, to advance society and promote unity. This sort of transracialism often exists in interpersonal relationships, and sometimes in whole communities, but on a national scale it falters. And even among the greatest champions of coalitions, there is room to diverge on the issue of Black power.

I put my reverse migration proposal to two coalition champions I highly respect: the Reverend Jesse Jackson and the Reverend William Barber. Jackson has dedicated his life to the building of a Rainbow Coalition and done so using his catchphrase "Keep hope alive!" He was frankly reticent about the idea, skeptical that people would migrate and convinced that Black people's greatest political "leverage" was still tied to racial alliances.

Barber is the father of the Moral Monday civil rights protests that began in North Carolina; in 2018 he reactivated the Poor People's Campaign, the multiracial project Martin Luther King was organizing when he was assassinated. As Barber pointed out, most of the people who marched with him in the Moral Monday protests were white. And yet he was open and agreeable to the issue of reverse migration. "From state up is the only way," he told me. "If you change the South, you change the entire nation." This is not surprising coming from Barber, whose own parents were reverse migrants who moved back south from Indiana to fight racism.

Barber illustrates that you can deeply believe in what he calls "fusion coalition" and cross-racial alliance and still support reverse migration. There is no contradiction in it.

The goal is to take bold action.

Black people are told that the racists will soon die off and we will be left with a society blissfully free of racial prejudice. This is magical thinking. Generations of people have died waiting on the racists to die out.

Black people are told that the racial progress America has made and continues to make is undeniable, and we should take heart in that fact.

It has been four hundred years since "20 and odd" enslaved Africans stolen from what would now be Angola disembarked the privateer *White Lion* at Point-Comfort, Virginia,

THE END OF HOPING AND WAITING

in 1619. The vast majority of that time has seen the active, vicious oppression of Black people in this country: 250 years of slavery, 100 years of Jim Crow, and now 50-plus years of mass incarceration.

And, every inch of so-called progress has come at staggering cost, demanding the shedding of tears and the shedding of blood.

Progress is the wall behind which white America hides. (Even many Black leaders have absorbed and regurgitate the progress narrative.) White liberals expect Black people to applaud their efforts. But how is that a fair and legitimate expectation? Slavery, white supremacy, and racism are horrid, man-made constructs that should never have existed in the first place. Are we meant to cheer the slow, creeping, centuries-long undoing of a thing that should never have been done?

Malcolm X was once asked if he felt that we were making progress in the country. He responded:

"No. I will never say that progress is being made. If you stick a knife in my back nine inches and pull it out six inches, there's no progress. If you pull it all the way out, that's not progress. The progress is healing the wound that the blow made."

I wouldn't even call that true progress. Even healed wounds remember trauma. In the 1989 *American Masters* documentary *James Baldwin: The Price of the Ticket*, Baldwin addresses the progress question this way:

> You've always told me that it takes time. It has taken my father's time, my mother's time. My uncle's time. My brothers'

and my sisters' time. My nieces' and my nephews' time. How much time do you want for your progress?

Inching toward inclusion is a luxury that the Black people cannot afford and therefore cannot abide. As Martin Luther King put it in his fiery "Letter from Birmingham Jail," the white moderate "lives by a mythical concept of time," one that for Black people translates into deferments and frustrations.

As I pondered Obama's hope speech that night in New York, it occurred to me that he was only the latest ambassador of the political hope doxology and the inevitable blindness that it renders, that he has many other comrades in that crusade, perhaps the most famous of whom being Booker T. Washington.

From Washington to Barack Obama—and beyond—Black leaders have preached a doctrine of hope, for freedom and equality as well as power, that has ultimately disappointed. In fact, I see Obama as a Washingtonian figure. There are differences between the two men, to be sure, but the similarities are striking. Both were young, brilliant biracial men, gifted orators, the most popular and important Black men of their age. Both men were viewed by white people—particularly white liberals—at the time of their emergence as remedies to the "race problem."

As Joe Biden said of Obama when he was campaigning in 2007, "You got the first mainstream African-American who is articulate and bright and clean and a nice-looking guy." That is how many white liberals saw Washington, a feel-good

THE END OF HOPING AND WAITING

Black leader, who would allow them to purchase absolution on the cheap.

But eventually both men were forced to confront the truth that all Black idealists must reckon with: White supremacy cannot be appeased. It can't be bargained with. It can't be convinced. White supremacy is ravenous and vicious. It is America's embryotic fluid. America was born in it and genetically coded by it. No amount of hoping or waiting, coalition-building or Kumbaya can redress that reality.

Racism is a flaw in the oppressor, not the oppressed. Viewing it to the contrary is a form of falling under the spell of a "politics of respectability," a term coined by the Harvard professor Evelyn Brooks Higginbotham.

In a way, Washington was the progenitor of respectability politics. His message fell softly, agreeably on white ears. He didn't demand, as much as he encouraged. He wasn't angry, he was amenable. He didn't insist upon atonement for the past, he focused on opportunities in the future. He wasn't seeking restorative justice from white people but simply offering an elevated class of laborer to them. His speeches offered assurances to white people and admonitions to Black ones.

This made Washington, born a slave, arguably the most influential Black person in the country, if not the world, at the turn of the twentieth century. And his soft-shoe celebrity increased not only his legend, but also his largesse—he began his life an enslaved child and ended it a renowned man who comfortably summered with his family in the Long Island hamlet of Fort Salonga in a charming two-story house on a bluff overlooking the sound.

And, as much as I find Washington's hope misplaced, I remain enthralled by his faith in it and his devotion to Black advancement, in part because his life repeatedly intersects with my own.

In April 1915 Washington arrived in Louisiana for an "educational tour" of the state. He had conducted these tours of several southern states, and each stop was a sensational event, throngs meeting him at train stations, and thousands, both Black and white, coming to hear him speak.

One of the last stops on his tour through Louisiana was a small college tucked away in the piney hills of the northern part of the state: Coleman College, a school established in 1885 to educate former slaves and their children, the same purpose for which Washington had established his Tuskegee Institute in Alabama just four years earlier.

Coleman was located in Gibsland, a town that was then, as it is now, composed of about a thousand people. The town was situated on land that had once been the plantation of a Dr. Jasper Gibbs, an enslaver.

Like Tuskegee, Coleman was supported in part by white philanthropists in the North, but also by the enthusiastic efforts of the Black people in the community. As a son of the school's founder once recalled, "Poor God-fearing people out of their meager holdings gave liberally; sometimes all they possessed. Individuals who had no money to contribute gave their time and labor to aid in erecting buildings."

A visit by Washington was seen by many as an anointing for a school or an institution, bolstering the credibility of their mission and vouching for the efficacy of their efforts, which in turn would allow for the raising of more money.

Coleman, like Tuskegee, was built by the students themselves, who even made their own bricks from the red clay on which the school rested and cured those bricks in a campus kiln. Coleman had been founded by a man named Oliver Lewis Coleman, who, much like Washington, was credited with using his institution to both better his race and produce racial harmony.

So, on that spring day, Washington stood on the school's campus, its impressive two- and three-story buildings, some brick, some wood, rising up from earth and hope, sitting astride a ridge one historian described as "nearly a complete horse shoe made by the hand of nature's God," that ridge providing the majestic effect of forcing all who approached to look up in awe at the building from the dirt road lined by rustic fencing. Washington was accompanied on the campus by the town's mayor, a white man by the last name of Lazarus, who extolled Coleman as he spoke:

> *There has never been any race trouble since Coleman came to this community. Coleman is to us all a guarantee of peace between the two races. Coleman has taken raw, gawky, unpromising country boys and made men of them.*

This statement was particularly important and noteworthy because it was delivered just two months after the release of the racially incendiary *Birth of a Nation*, a bona fide blockbuster. The film would spur the resurgence of the Klan, its membership soaring to the millions by the mid-1920s. There was a screening of the film at the White House during which President Woodrow Wilson is widely (and possibly apocryphally) cited as having declared the film "like history written with lightning, and my only regret is that it is all so terribly true."

Wilson, a racist through and through, was the first southern-born president since the Civil War. And yet, as the historian Kenneth O'Reilly has written, Wilson "was the first Democratic presidential candidate in history to receive widespread endorsement from prominent Blacks" by promising that he would be fair, and even "W. E. B. Du Bois considered him 'a cultivated scholar' of 'farsighted fairness'" who "will not seek further means of 'Jim Crow' insult."

To the contrary, Wilson brought Jim Crow to the federal government, institutionalizing segregation within the federal civil service. Du Bois would be bitterly disappointed.

The Louisiana tour would be Washington's last, as he died a few months later, and Coleman would be one of his last stops on that tour.

Coleman College would become the high school I attended some one hundred years later, Gibsland-Coleman High School, still located on the same majestic ridge, still laboring to hold the peace, still toiling to turn gawky country children into productive adults.

And, it had largely worked. The racial violence that scarred much of the rural South largely skipped over our town. And the school imbued in its students a particular form of pride, one born of heritage and history, one that straightened the back and adjusted the gait. There was something special in the comportment of "Coleman kids."

It certainly had an effect on me. I thought often about the

THE END OF HOPING AND WAITING

profound legacy of walking the same ground for the same reason that these children of the formerly enslaved had done a century before. I stared at the pictures of the old building, now lost to the ages, particularly the three-story Victorian girls' dormitory with its arched entrances and soaring belfry, built by the students from the bricks they made. I was struck also by the image of the faculty, perfect in posture, determined in demeanor, the same faculty who chose to forgo their pay when times got hard, so that the school could stay open.

This was a place suffused with the spirit of self-sufficiency and self-determination, built by Black hands and guided by Black souls.

And, the fact that Washington had endorsed it endeared him to me in a personal way.

My great-great-grandfather William Blow, the oldest ancestor for whom I have found records, was born in Alabama around 1850, which made him a few years older than Washington. He was born during slavery, but not necessarily enslaved. According to one genealogical record of my home parish of Bienville, Louisiana, "Paternal ancestor of William said to have saved money for many years to buy his freedom."

It is the family's belief that an ancestor was brought to Alabama with an enslaver, Peter Blow, who moved in 1818 from Southampton, Virginia, to Alabama with a handful of slaves, including Dred Scott, who was once called Sam Blow. Peter Blow and his enslaved remained in Alabama for twelve years before moving in 1830 to Missouri. One tantalizing line in that Bienville Parish record seems to support our family's theory: "A Benjamin Blow, free man of color living in Southampton Co. VA in 1850 has been put forth as an ancestor." As far back as I can trace, my patrilineality consisted of free Black men living in the South.

It should be noted that Southampton is the same county in which Nat Turner was born, the same year or shortly after Dred Scott, and where Turner staged his rebellion in 1831.

William's family was pass-for-white light, not dissimilar in appearance from Washington himself. Washington was the child of a Black mother, enslaved, and an unknown white man; there is no record of a white man or woman in my family's records that I've found. Like most mixed-race children born during slavery, their very existence was often evidence of a crime, one through which the perpetrator then conferred a privilege.

Like Washington, my ancestors' lineage in Alabama isolated them from the bulk of formal benefits—legal, economic, and cultural—that transmit to white people as birthright, but informally they reaped lesser rewards because whiteness would still nod to whiteness, even in cases where it was thought diluted and adulterated, even when bastardized by Blackness. To many, being part white was still better than being all Black, especially when considered through the white gaze. Even Black people, drowning in the doctrine and practice of white supremacy, would elevate and ordain light-skin privilege, white privilege's residue.

As the historian Michael W. Fitzgerald wrote in his 2017 book, *Reconstruction in Alabama*, mixed-raced Blacks in the state were more urban, better educated, and more well off before and after the Civil War. "Color distinctions mattered:

THE END OF HOPING AND WAITING

in 1860, seventy-five of the eighty wealthiest free black families in Mobile were headed by 'mulattoes.'"

My ancestors were by all accounts relatively well off. William owned one hundred acres of land in Elmore County that he farmed, less than forty miles from Washington's Tuskegee Institute. He had twenty-two children, eleven by his first wife and eleven by his second. There is no way of knowing if the two men knew each other or ever met, but William was apparently so taken with Washington that he named his twentieth child, born in 1908, Booker T.

I never heard my family or townsfolk speak of Washington, and yet his philosophy seemed to suffuse our very existence. Learn a trade, work hard, ruffle few feathers.

I spent the first three years of my life living with my grandmother in Arkansas as my middle brother had done before me. I eventually came home. My middle brother never did. My mother had five boys over eight years; I was the last. With every pregnancy, for reasons unknown, she became horribly ill. My middle brother Marvin was born just nine months after the second brother, Frank. My grandmother stepped in to ease the burden until my mother recovered by taking Marvin to live with her. But the months and years slowly stretched into an entire childhood and he became more attached to the grandmother who did the raising than to the mother who did the birthing.

They lived in a small, all-Black community just miles from the Louisiana border. It had been homesteaded in 1866 when the plantation owner who owned it died, and his son gave the land to his former slaves, who had been purchased in New Orleans in 1841. This is one of the great ironies of slavery: sometimes the greatest allies in the advancement of southern Blacks were southern whites who had been children of the enslavers. As W. E. B. Du Bois wrote in *The Souls of Black Folk*:

> To-day even the attitude of the Southern whites toward the Blacks is not, as so many assume, in all cases the same; the ignorant Southerner hates the Negro, the workingmen fear his competition, the money-makers wish to use him as a laborer, some of the educated see a menace in his upward development, while others—usually the sons of the masters—wish to help him to rise.

It was the sons of the man who enslaved Dred Scott, men who had grown up with Scott and became friends of his, who helped pay his legal fees in the *Dred Scott v. Sandford* case of 1857 and who bought his freedom when the Supreme Court ruled against him in one of the worst decisions the court has ever rendered.

When we visited Kiblah we often played in the old Negro school that had since been converted to a community center, unaware that the building was a Rosenwald School that would eventually be placed on the National Register of Historic Places. As part of a plan developed by Washington and his friend Julius Rosenwald—the president of Sears, Roebuck and Co. and one of the richest men in America at the time—to build quality schools for Black children in the segregated South, nearly five thousand of these schools were built, constituting possibly the most important instrument of Black education in the twentieth century. According to the

THE END OF HOPING AND WAITING

National Trust for Historic Preservation: "By 1928 one in every five rural schools in the South was a Rosenwald school; these schools housed one third of the region's rural black schoolchildren and teachers. At the program's conclusion in 1932, it had produced 4,977 schools, 217 teachers' homes, and 163 shop buildings that served 663,625 students in 15 states."

Only about 10 to 12 percent of those structures survived, and they have collectively been added to the National Trust's list of America's 11 Most Endangered Historic Places.

To this day my best friend is a man named Booker T. Washington, who went to college with me and is my daughter's official godfather and all my children's spiritual one.

It is because I felt so familiar with Washington, felt such an affinity for his message of self-reliance and hard work, that it was always impossible for me to comprehend his political accommodation of and acquiescence to white power.

Washington's 1895 Atlanta Exposition address rightly stands as a seminal moment in the African-American power narrative, one in which Black people's greatest leader assuaged white people's greatest fear, and thereby gave a casual stamp of approval to white supremacy in affairs of state.

Washington was hesitantly chosen by the white promoters of the Atlanta Cotton States and International Exposition to address a mostly white crowd as illustration of the racial progress the South had made since the Civil War. He didn't disappoint. He insisted that the Negro would have to earn his rightful place rather than insisting, as the writers of the Declaration of Independence put it:

that all men are created equal, that they are endowed by their Creator with certain unalienable Rights, that among these are Life, Liberty and the pursuit of Happiness.—That to secure these rights, Governments are instituted among Men, deriving their just powers from the consent of the governed.

Instead, as Washington told his audience:

The wisest among my race understand that the agitation of questions of social equality is the extremest folly, and that progress in the enjoyment of all the privileges that will come to us must be the result of severe and constant struggle rather than of artificial forcing. No race that has anything to contribute to the markets of the world is long in any degree ostracized. It is important and right that all privileges of the law be ours, but it is vastly more important that we be prepared for the exercise of these privileges. The opportunity to earn a dollar in a factory just now is worth infinitely more than the opportunity to spend a dollar in an opera-house.

He went on to assert that "in all things that are purely social we can be as separate as the fingers, yet one as the hand in all things essential to mutual progress."

But who then sits in judgment of the fulfillment of the Negro's "severe and constant struggle"? The white men in power. Washington seemed to believe that it was possible—or necessary—for Black people to earn their way into the white man's equality, that for Blacks "it is at the bottom of life we must begin, and not at the top."

Yet history has shown us that white supremacy will not be

moved by the bowing of the Black back, but by the breaking of it.

I grieve both for Washington and for Black people. I believe fully that Washington wanted only the best for Black people and that he labored his entire life to secure it, but his position on forgoing political power to focus on economic stability was one of the greatest miscalculations in African-American history, and he proposed it at a time when the Black population in America was at its most concentrated and therefore potentially on the verge of its greatest political strength.

It seems to me that Washington suffered a form of Stockholm syndrome, having fallen somewhat in love with the culture that held his people in captivity. As he put it in his 1901 autobiography, "I have long since ceased to cherish any spirit of bitterness against the Southern white people on account of the enslavement of my race." This sentiment, and the overall spirit of Washington's exhibition address, suffused Obama's Philadelphia race speech of 2008, considered by many to be one of his most consequential. Both speeches were aimed at calming and assuring white people, and Obama, in his own way, implored his audience to move beyond their anger, asserting that it "prevents the African-American community from forging the alliances it needs to bring about real change." But, he also went on to draw a harmful false equivalency between Black people's anger over centuries of anti-Black, white supremacist terror, and white people's anxiety in response to the relatively recent phenomena of economic stagnation and displacement, affirmative action and crime: in his words, "a similar anger exists within segments of the white community." He elaborated:

They've worked hard all their lives, many times only to see their jobs shipped overseas or their pensions dumped after a lifetime of labor. They are anxious about their futures, and they feel their dreams slipping away. And in an era of stagnant wages and global competition, opportunity comes to be seen as a zero sum game, in which your dreams come at my expense. So when they are told to bus their children to a school across town; when they hear an African-American is getting an advantage in landing a good job or a spot in a good college because of an injustice that they themselves never committed; when they're told that their fears about crime in urban neighborhoods are somehow prejudiced, resentment builds over time.

But of course, Washington went even further than Obama. He considered the Black people who "went through the school of American slavery" to be "in a stronger and more hopeful condition, materially, intellectually, morally and religiously, than is true of an equal number of black people in any other portion of the globe." As he saw it: "notwithstanding the cruel wrongs inflicted upon us, the black man got nearly as much out of slavery as the white man did."

Washington's hero and mentor was a white man, General Samuel C. Armstrong, a former Union army officer who founded Hampton Normal and Agricultural Institute for the education of Black people in Hampton, Virginia, in 1868. The school is now called Hampton University.

But whatever Armstrong's evolution on race may have been, this was a man who was born to Hawaiian missionaries and whose first introduction to Black people came during the

THE END OF HOPING AND WAITING

Civil War. His impression wasn't charitable, as he wrote to a friend from Williams College in 1862:

> *Chum, I am a sort of abolitionist, but I haven't learned to love the Negro. I believe in universal freedom; I believe the whole world cannot buy a single soul. The Almighty has set, or rather limited, the price of one man, and until worlds can be paid for a single Negro I don't believe in selling or buying them. I go in, then, for freeing them more on account of their souls than their bodies, I assure you.*

Yet, Washington would refer to Armstrong as "a great man—the noblest, rarest human being," and the "strongest, most beautiful character" he had ever met.

Armstrong's position speaks to the insidiousness of white supremacy, a concept that many try to apply only to vocal, violent racists, but which is in fact more broadly applicable and pervasive. People who want to help you as well as those who openly hate you can be white supremacists.

It is possible to promote the uplift and education of Black people and simultaneously believe that the highest station attainable to Black people, as a whole, is well below that of whites. You could be in favor of abolition—and even risk your life in battle for that cause—and still be a white supremacist.

People think that they avoid the appellation because they do not openly hate. But hate is not a requirement of white supremacy. Just because one abhors violence and cruelty doesn't mean that one truly believes that all people are equal—culturally, intellectually, creatively, morally.

Entertaining the notion of imbalance—that white people are inherently better than others in any way—is also white supremacy. This view—that Black people are not human at the same level as white people—is what King called "the thing-ification of the Negro." It is passive white supremacy, soft white supremacy, the kind divorced from hatred. It is permissible because it's inconspicuous. But this soft white supremacy is deadlier, exponentially, than demonstrative racism. Indeed, from the beginning, even many liberals who opposed slavery *in no way* believed in Black equality. They too were white supremacists.

In Washington's case, his white friendships seemed to render him blind—willfully and stubbornly—to the raging white supremacy encircling him. Writing about the friends at Tuskegee, he opined: "The Tuskegee school at the present time has no warmer and more enthusiastic friends anywhere than it has among the white citizens of Tuskegee and throughout the state of Alabama and indeed the entire South."

The very year Washington wrote this, racist politicians in Alabama called a constitutional convention with the express purpose of stripping Black people of the right to vote and of canonizing white supremacy.

As the *Journal of Negro History* detailed in a 1949 article titled "Populism and Disfranchisement in Alabama":

> *The Democratic State Executive Committee met in Montgomery on April 19 for the purpose of getting reports from the field and to brief candidates for delegates to the proposed convention. Emmet O'Neal, later to become governor of the state, stated that "the paramount purpose of the constitutional*

> convention is to lay deep and strong and permanent in the fundamental law of the State the foundation of white supremacy forever in Alabama, and that we ought to go before the people on that issue and not suggest other questions on which we differ." Candidate Thomas J. Long, from Walker County, reminded his fellow candidates that "the way to win the fight is to go to the mountain counties and talk white supremacy. . . . I don't believe it is good policy to go up in the hills and tell them that Booker Washington or Councill or anybody else is allowed to vote because they are educated. The minute you do that every white man who is not educated is disfranchised on the same proposition."

Washington would note in his memoir, *Up from Slavery*, "It is a pleasure for me to add that in all my contact with the white people of the South I have never received a single personal insult." Assuming this assertion is accurate and honest, I would wager that not many Black people at the dawn of the twentieth century could make such a boast.

The year Washington wrote this, he was invited to dinner at the White House by President Theodore Roosevelt. The response was swift and vicious, producing a great hue and cry from the southern white press, so much so that the White House tried to redefine the occasion as not a dinner but a luncheon. After the dinner, Senator James K. Vardaman of Mississippi asserted that the White House was now "so saturated with the odor of nigger that the rats had taken refuge in the stable."

As Vardaman would put it in 1904: "I am opposed to the nigger's voting, it matters not what his advertised moral and mental qualifications may be. I am just as much opposed to Booker Washington as I am to voting by the cocoanut-headed, chocolate colored typical little coon, Andy Dotson, who blacks my shoes every morning. Neither one is fit to perform the supreme functions of citizenship."

Vardaman, one of the most racist politicians in American history, would go on to become governor of Mississippi, and would famously state during a speech in Poplarville, Mississippi, "If it is necessary, every Negro in the state will be lynched."

I have tried to consider Washington's accommodation in light of the precarious position in which the formerly enslaved found themselves. They had been freed into enemy territory without land, education, health care, or a way to make a living.

As the Reverend Dr. Martin Luther King Jr. put it in his "Other America" speech at Stanford University in 1967:

In 1863 the Negro was freed from the bondage of physical slavery. But at the same time, the nation refused to give him land to make that freedom meaningful. And at that same period America was giving millions of acres of land in the West and the Midwest, which meant that America was willing to undergird its white peasants from Europe with an economic floor that would make it possible to grow and develop, and refused to give that economic floor to its black peasants, so to speak.

In another speech, King extended his complaint:

THE END OF HOPING AND WAITING

But not only did they give the land, they built land-grant colleges, with government money, to teach them how to farm. Not only that, they provided county agents to further their expertise in farming. Not only that, they provided low interest rates in order that they could mechanize their farm. Not only that, today many of these people are receiving millions of dollars in federal subsidies not to farm. And, they are the very people telling the black man that he ought to lift himself by his own bootstraps.

At Stanford, King told the assemblage:

This is why Frederick Douglass could say that emancipation for the Negro was freedom to hunger, freedom to the winds and rains of heaven, freedom without roofs to cover their heads. He went on to say that it was freedom without bread to eat, freedom without land to cultivate. It was freedom and famine at the same time.

Indeed, the enslaved were freed into a public health catastrophe. Connecticut College history professor Jim Downs explained in the *Lancet* in 2012 that bickering among federal, state, and city officials over who was ultimately responsible for providing health care for the newly freed "created an institutional vacuum that left ex-slaves defenseless against disease outbreaks, and their situation was further exacerbated by freed-people's nebulous political and economic status." Downs estimates that a quarter of the formerly enslaved got sick or died between 1862 and 1870.

Through this lens we can follow Washington's logic in addressing this social epidemic of economic suffering, starvation, and death, before attending to political imperatives. But, the dire conditions in which Black people found themselves was a direct result of oppression, and oppression is enabled—or alleviated—by political power.

W. E. B. Du Bois, another towering figure among spokesmen for Black America, initially praised Washington's Atlanta speech, writing to him in a letter, "Let me heartily congratulate you upon your phenomenal success at Atlanta—it was a word fitly spoken."

But, upon further consideration, Du Bois, like many other leading Black intellectuals, came to see the speech as a tragic mistake. In his 1903 *The Souls of Black Folk*, he issues a stinging rebuke to Washington, dubbing his speech "The Atlanta Compromise." In the essay "Of Mr. Booker T. Washington and Others," Du Bois writes that Washington "represents in Negro thought the old attitude of adjustment and submission," and that his proposal "practically accepts the alleged inferiority of the Negro race."

It was right that Du Bois would condemn Washington's political submission and accommodation of white racists in the South.

And Du Bois was right to assert that "Negroes must insist continually, in season and out of season, that voting is necessary to modern manhood, that color discrimination is barbarism, and that Black boys need education as well as white boys."

As Du Bois pointed out, the reward for accommodation is subjugation. He wrote, "As a result of this tender of palm-branch, what has been the return?" pointing out that in the years since Washington's speech, Black America had suffered disenfranchisement, "the legal creation of a distinct status of civil inferiority for the negro," and a dwindling aid for Black colleges.

This is all true, but some of Du Bois's own theories were problematic. His adoption and promulgation of the Talented Tenth concept—a theory developed by a white man, Henry Lyman Morehouse, for whom the historically Black Morehouse College was named—was utterly elitist and self-interested, favoring men like Du Bois himself.

The theory champions a sort of trickle-down equality in which the anointed and endowed plow the way for the lesser. But, that is not the way aristocracies—of any sort, of any group—operate. The aristocracy's primary motivation is the maintenance of advantage and the benefits that accrue from it. Benevolence may be peripheral, but it's by no means central.

Too many of the Black elite get drafted into a white-adjacent privilege, suckled by personal prosperity and personal comfort, blinded by the glamour of the high society. They become the neo house Negroes, placated, passive, a resurrection of an antebellum relic in which the best and brightest of Black society, those who would otherwise be the generals in resistance and rebellion, are lulled to sleep by luxuries.

The more talented and successful you are, the more tightly the moneyed establishment embraces you, cleaves you from the struggling plight of your people, and beknights you as an honorary member of theirs. It is easy to get lost in this, seduced by it, convinced of it.

I know this world well. I could easily have spent the rest of my life filling my calendar with Park Avenue parties, exclusive salons, and destination vacations. I was part of what the writer Holly Peterson calls New York's "Accomplisher Class," in which "what people admire is the top achievement in almost any field." Du Bois too was part of this class.

Many of these people greeted my decision to move back south with bewilderment, treating it as a betrayal if not an insanity. How could a Black man, having risen to the height of New York's white cultural inclusion, spurn it?

Easily. That inclusion offered only personal comforts, not community recovery. The elite class, neither Black nor white, liberal or otherwise, will never offer a path of restitution and restoration for the Black masses. Black healing and rebuilding will come from the bottom up, not the top down.

And, there is something disconcerting and off-putting about this northern man, Du Bois, born free after slavery into the relative comforts of a small, majority-white Massachusetts town, where he was reportedly treated well and went to an integrated school, lecturing a man born into slavery and who struggled with the brutality of the South where 90 percent of Black people lived.

Du Bois's words drip with condescension. As Du Bois himself would put it, "Mr. Washington knew the heart of the South from birth and training." Du Bois did not. The only time he spent in the South was to attend Fisk University in Nashville and then as a professor at Atlanta University, from where he was summarily dismissed by the school's president Rufus Clement.

As Isabel Wilkerson writes in *The Warmth of Other Suns*:

THE END OF HOPING AND WAITING

Clement would be at odds with Du Bois almost from the start, perhaps threatened by the long shadow of his celebrity or put off by the elder man's impertinent disregard for Clement, who was thirty years younger than Du Bois. But it was just as likely a contest between the accommodating pragmatism of the southern-born Clement and the impatient radicalism of the northern-bred Du Bois. The two men were the very embodiment of the North-South divide among black intellectuals.

This split manifested by Washington and Du Bois has presented in every generation. When the South gave us the Reverend Dr. Martin Luther King Jr., John Lewis, and Jesse Jackson, the destination cities gave us Malcolm X, Huey Newton, and Stokely Carmichael. When the South gave us the Southern Christian Leadership Conference and the Student Nonviolent Coordinating Committee, the destination cities gave us the Black Panthers and the Nation of Islam. The severing point is often whether you believe, as King did, that America is in need of redemption and capable of it—if you believe that America is just sick, and needs to be healed—or if you believe that what ails America on the racial question is terminal and hopeless.

I have a foot in both camps. I have a particular faith in the capacity of this country to grow and to heal, but I also doubt the country can ever be fully, truly cured of racism. I do believe that the victims of racism will eventually, inevitably, prevail over the racists, and they will do so as an act of strategy, not persuasion.

The most important and effective moment for African-Americans, the civil rights movement, was born of the people who stayed in the South. Black people in northern destination cities added tremendously and disproportionately to the art, culture, and literature of Black America; the children of the Great Migration have made poetry of their pain. But the Blacks who remained in the South waged the lion's share of the war.

The split is perfectly illustrated by a story in a scene from Stanley Nelson's fascinating documentary *Freedom Summer*. When an unofficial, integrated group of delegates from Mississippi claimed the seats of the official all-white one at the 1964 Democratic convention, the party struggled to figure out a way to seat one of the two groups of delegates with the least amount of turmoil. At one point, the storied Black congressman Adam Clayton Powell Jr. of New York was dispatched to the integrated delegation to persuade them to accept a compromise arrangement—the seating of the all-white delegation with the addition of just two members of the integrated delegation—to end the standoff.

Powell reportedly said to Fannie Lou Hamer, a member of the integrated delegation, "You don't know who I am, do you?" Hamer responded, "Yeah, I know who you are. You are Adam Clayton Powell." She continued, "But how many bales of cotton have you picked? How many beatings have you taken?"

Washington himself was a product of experience. And yet he displayed a remarkable blindness to the white supremacist terrorism playing out all around him as he sought to paint a world as he hoped it to be rather than the way it actually was. He would write in his 1901 memoir:

THE END OF HOPING AND WAITING

I have referred to this unpleasant part of the history of the South simply for the purpose of calling attention to the great change that has taken place since the days of the "Ku Klux." Today there are no such organizations in the South, and the fact that such ever existed is almost forgotten by both races. There are few places in the South now where public sentiment would permit such organizations.

At least fourteen people would be lynched in Alabama the year Washington wrote these words and over one hundred more would be lynched in subsequent years, as hundreds had been lynched there in preceding years.

Washington's theories, though admittedly consequential, were in many ways philosophical; there was another Black man, also guilty of pacifying white supremacy, who arguably had an even greater impact in terms of policy. His name was Isaiah T. Montgomery, the lone Black delegate to the Mississippi constitutional convention that took place five years before Washington's Atlanta speech.

As the historian, sociologist, and author James W. Loewen has written, the Confederates "won the Civil War in 1890." That was the year that Mississippi passed its new constitution at a convention specifically convened to write Black voter suppression into the heart of the state. Loewen explained:

As one delegate put it, "Let's tell the truth if it bursts the bottom of the Universe. We came here to exclude the Negro. Nothing short of this will answer." The key provision to do so was Section 244, requiring that voters must be able to give a "reasonable interpretation" of any section of the state constitution. White registrars would judge "reasonable." Other states across the South copied what came to be called "the Mississippi Plan," including Oklahoma by 1907.

That delegate was Solomon Saladin Calhoon, president of the convention, who would go on to become one of the three judges on the Supreme Court of Mississippi. In the months leading up to the convention he would write that "Negro suffrage is an evil, and an evil to both races. Its necessary outcome is that conflicting aspirations and apprehensions must produce continual jars and frequent hostile collision, which do not occur with homogeneous races." Calhoon called for the clear and immediate elimination of that evil. He even surfaced the idea of his own form of racial colonization of the South, so afraid was he of the Black majorities. He wrote:

> *It has been said that the cure for the evil will appear in white immigration in the Black belt which shall furnish a preponderance of our own race. The introduction of whites into the South is very desirable, and in mitigation of the evil. The South is keenly alive to the importance of it and is encouraging it in every way she can.*

But colonization wasn't quick enough for their purposes. During the morning session on the fortieth day of the Mississippi constitutional convention of 1890, delegate George P. Melchior, a Bolivar County planter, put it: "It is the manifest intention of this Convention to secure to the State of Mississippi, 'white supremacy.'"

At the convention, Montgomery was placed on the elective franchise, from which he passionately advocated for a literacy test at the polls, explaining that he knew well that it would overwhelmingly disenfranchise Black voters who were then the majority in Mississippi. As he outlined in his infamous "Peace Bush" speech:

> This bill will affect the voting population of the State as follows: Present voters, white 118,890; this bill will restrict 11,889, and leave a net white vote of 107,001. Present voters, negroes 189,884; this bill will restrict 123,334; and leave a negro vote of 66,550, giving a white majority of 40,451.

Montgomery acknowledged that he knew many of the people he would be disenfranchising personally and that their "hearts are true and steel," and that these ranks included many who were "soldiers who had stood amid the smoke of battle on bloody field in defense of the flag." And yet, he submitted,

> I wish to tell them that the sacrifice has been made to restore confidence, the great missing link between the two races, to restore honesty and purity to the ballot-box and to confer the great boon of political liberty upon the Commonwealth of Mississippi. Mr. President, I wish to be distinctly understood in saying that there are terms upon which I consent to lay the suffrage of 123,000 of my fellow-men at the feet of this convention.

Other southern states soon followed Mississippi's lead, calling constitutional conventions of their own and instituting a raft of Jim Crow policies, including the poll test that Montgomery championed, and he would apparently come to regret his decision at the convention.

Montgomery is sometimes called Mississippi's Booker T. Washington, but I don't think the comparison fair or accurate. Not even Washington made a miscalculation of this magnitude that would directly impact Black electoral power.

As Frederick Douglass said of Montgomery at the time: "He has made peace with the lion by allowing himself to be swallowed."

For 150 years, Black Americans have been hoping and waiting. We have marched and resisted. Many of our most prominent leaders have appeased and kowtowed. We have seen our hard-earned gains eroded by an evolving, refining white supremacy, while at the same time we are told that true and full equality is in the offing. But, there is no more guarantee of that today than there was a century ago.

In Washington's Atlanta speech he implored Black people in the South "who depend on bettering their condition in a foreign land or who underestimate the importance of cultivating friendly relations with the Southern white man, who is their next-door neighbor . . . [to] cast down your bucket where you are" and make friends with your oppressors.

I say to Black people, return to the South, cast down your

THE END OF HOPING AND WAITING

anchor there, and create an environment in which oppression has no place.

As Frederick Douglass once wrote about escaping slavery, "I prayed for twenty years but received no answer until I prayed with my legs."

Black people must pray with their legs.

SIX

THE REUNION

❈ ❈ ❈

BLACK POWER SIMPLY MEANS: LOOK AT ME, I'M HERE. I HAVE DIGNITY. I HAVE PRIDE. I HAVE ROOTS. I INSIST, I DEMAND THAT I PARTICIPATE IN THOSE DECISIONS THAT AFFECT MY LIFE AND THE LIVES OF MY CHILDREN. IT MEANS THAT I AM SOMEBODY.

—Whitney M. Young in *Jet* magazine in 1971

On a cold and dreary morning I sat talking with Samaria Rice in a rental car in Cleveland just yards away from the gazebo where one year prior, to the day, a policeman's bullet ripped its way through the abdomen of her twelve-year-old son, Tamir.

She vacillated between cheer and despair, a phenomenon I had observed with other so-called Mothers of the Movement, those Black women whose children had been killed by police, catapulting them into activism when the natural impulse was to withdraw and weep. When we met a few hours earlier she confessed how emotionally spent she was: "I'm tired and I'm overwhelmed, and I just want to go to bed."

Her words echoed those of another mother, Audrey DuBose, whose son Sam was shot in the head by police in Cincinnati just four hours away. After I followed Audrey to a television interview and watched a performance honed by repetition and necessity, she confessed in a whisper: "All I want to do is just shut my door and cover up and never open it again."

This pull to withdraw, to cover up, to curl up, was always at war with their desire—and the culture's desire for them—to be a champion for the slaughtered child. During my first in-person interview with Sybrina Fulton, the mother of Trayvon Martin, one of the things that struck me most was the reflexive, unconscious way she wrapped her hands around her mother's arm and rested her head on her shoulder like a discomforted child, even as she dutifully answered my questions and fought for the memory of her son.

There was a gaping hole in these women, a void, as each was inducted into the sorority of sorrow.

Over the years of interviewing these mothers, I came to a greater appreciation of war correspondents, their ability to witness all manner of personal loss and grief and yet retain enough emotional distance to chronicle it, but not so much distance as to become numb to it. And make no mistake: The police killings were products of a war—asymmetric, government-sanctioned, and unremitting. The killings were the collateral damage, the logical extension of the criminalization of Blackness, the militarization of policing, and the commodification of penalty.

Before Samaria showed me where the police cruiser had barreled across the small park near the Cudell Recreation Center, before she showed me how short a distance the gazebo was from her home in the three-unit brick apartment building off Madison Avenue, before she pointed to the spot where her son bled out onto the snow-dappled earth, she bared her soul to me, painfully, unflinchingly, confiding how difficult her own life had been.

Her early memories were of drugs and violence in her home. When she was twelve, her mother killed her father. She was compelled to testify at the trial. Her mother went to prison for manslaughter and Samaria was shuttled into the foster care system, experiencing a string of caregivers unfamiliar with giving and caring.

THE REUNION

As Samaria would tell the columnist Connie Schultz, "This family has always had parts of it that were destroyed."

She talked about her own run-ins with the law, for drugs and violence. And she talked about how she felt that the society around her had failed her, how all the power supposedly there to protect her, advocate for her, respond to her, felt corrupt.

There was a small ceremony at the gazebo to mark the day. Attendees were sparse, but news crews were plentiful. Samaria transformed herself when it came time to address the crowd, as I had seen other Mothers of the Movement do, into a social justice heroine, fierce mother of a slain child, tireless truth-teller confronting a predatory system. When it was over, we got back into my car, and she deflated, once again becoming the broken woman I'd met that morning.

When I interviewed her again the following year, she told me that she had Tamir's body cremated. His urn, a white box with blue marbling, rests on a shelf in her dining room among a collection of his toys—a couple of brown teddy bears; two Matchbox cars, one green, the other yellow; a miniature motorcycle; a Cleveland Cavaliers hat with the size sticker still affixed. I asked why she chose cremation. "When I leave, I don't want to leave him in Cleveland," she said.

I felt so close to Samaria each time we spoke, the way I had felt with the other mothers. Most, like me, had been raised poor or working class. Most, like me, had struggled. And it seemed that they could see in me what I could see in them: familiarity and communion. I believe it is one of the reasons they all dropped their guard so quickly. I wasn't there to exploit them or to judge; I was there because I recognized them. I knew them. I could *see* them.

What was supposed to be a thirty-minute interview turned into an all-day affair. When I finally excused myself and said goodbye at a dinner she insisted I attend, it was nine p.m.

And yet as I thought about the day I had spent with Samaria I couldn't help but consider just how different our formative experiences had been—hers in a destination city, and mine in the South.

Samaria and her family didn't live on the east side of Cleveland where most Black people lived, but on the north side. While the city of Cleveland is majority Black and about a third white, Samaria's Cudell community was a third Black, nearly 40 percent white, and nearly a quarter Hispanic, racially ideal, on paper, but far from idyllic. It was plagued by crime and violence.

We had both grown up poor, and we were not that far apart in age. And yet we had experienced poverty—and life—in such drastically different ways. Sure, there was violence in my upbringing. Poverty is almost always attended by violence. And as a child, my mother had even shot at my father. But unlike in Samaria's case, I don't believe it was my mother's intention to cut him down, but to scare him off. None of the bullets found their mark. She had let him back in that night, but the next morning it was made painfully clear that he hadn't changed his ways when a small child called my mother's phone, asked to speak to my dad, and told him his mother wanted him to meet her down the road at the neighborhood juke joint.

But I had never felt a threat from the state. And I could never have imagined anyone in my neighborhood perceiving me as a threat of any kind, let alone calling the police on me.

THE REUNION

Like most boys in my hometown, I'd had guns from the time I could walk, graduating from water guns and cap guns to BB guns and pellet guns to actual rifles and shotguns. It was nothing to see a pack of boys walking the streets with these weapons. Everyone who saw us knew us. They simply smiled and waved.

Our little town had only one police officer. When I was young, it was a kindly old man with obsidian skin and flour-white hair named Mr. Pero. As a teenager, it was my cousin by marriage, Mr. Cato. Even when I went to college, just twenty minutes away, the officers, all of them Black, seemed to know me or know of my family.

I had never known what it truly felt like to have the power of the state arrayed against me. I always felt safe in my majority-Black hometown and my majority-Black college town. Even the white officer who had threatened me and my friend in the majority-white town when I was in college felt to me more anomalous than representative. I never understood how much of a gift that was until I ventured north and that sense of safety was replaced by a stalking sense of dread.

I wanted for Samaria what I'd had for myself: a sense of security and control, a community that nurtured and nourished you, a government beholden to you and responsive to you. In fact, I wanted it again for myself. And, I believed that we could both have it, in the South.

In 2019, Janean, a friend of thirty years, invited me to inspect the new house she was building in the exurbs of Atlanta, some thirty minutes due south of the city.

THE DEVIL YOU KNOW

We had met in college. We were so much alike: both dark brown country kids raised by single parents—her by her father and me by my mother. In fact, she seemed to me my mother in miniature, and I reminded her of her father, so we formed an immediate, instinctual familial connection. We have called each other brother and sister ever since, doing it so consistently that it has sometimes caused amused confusion for those none the wiser.

We had both grown up poor. More precisely, I grew up among the "near poor," the demipoor, the "professional but struggling," while she had experienced poverty in the extreme.

As a grown woman with a gaggle of boys, my mother had gone to night school after work as a poultry processor and then a teacher's aide to earn her college degree and become a home economics teacher, a job to which she had long aspired. Teaching was one of the only professions in town, one of the only ones requiring a degree, and teachers were esteemed for that status and achievement.

But, my mother was a single mother with five children (one who lived mostly with my grandmother) and an elderly uncle living with us, for whom she was caretaker. With such a large family, we existed just a breath above the poverty line. My mother's first teaching job in 1975 earned her about $8,400 a year, or $525 in take-home pay a month. The median family income that year for a family our size was $14,600.

Janean's family circumstances had been more dire in Halley, Arkansas, a tiny town of fewer than four hundred people just west of the Mississippi River in the southeastern corner of the state.

THE REUNION

Since the sixth grade she had lived with her father and brother in a two-bedroom trailer with no running water, off a dirt road, dirt poor. When she was accepted into college in 1988, her father earned about $800 a month fixing tractors at a John Deere affiliate. Her father borrowed a company truck, drove her to school, and dropped her off with only a couple hundred dollars in her pocket.

She arrived not even realizing that she needed to pay tuition. When she was told she had to pay more than $4,000, she collapsed in tears. She didn't have it and knew no way to get it. A financial aid officer calmed her and walked her through the applications for a Pell Grant and loans. So began her college career.

My mother earned about $1,300 a month my freshman year, and I'd earned scholarships. I was also still eligible for a Pell Grant. This meant that the same financial aid office that came to Janean's rescue largely with borrowed money issued me a refund of a couple thousand dollars, because my aid was greater than my debt.

We had both struggled, but not evenly, and our matriculations would not be even. Mine was advantaged. There were layers at the bottom of our barrel.

The imbalance persisted after college. My first job was to produce maps, charts, and diagrams at the *Detroit News*. Hers was as an assistant manager at a Wal-Mart in Shreveport. She bounced from retailer to retailer, Lane Bryant to Dillard's, before she entered banking as a teller.

I would eventually rise to become a columnist, television commentator, and author. She would rise to oversee all the SunTrust banking centers in the state of Maryland and a few in northern Virginia. Now she was back in Atlanta as regional district manager for a different bank.

Both our journeys had been improbable, and we had traveled the road together, cheering and crying together, celebrating every promotion and accomplishment while consoling each other through our darkest days, casting no judgments, displaying no jealousies.

Now she was overflowing with excitement because she was building her dream, a five-bedroom, four-bathroom house in a country club community, off a country road, with a faux waterfall at the entrance. According to her, the development was about 80 percent Black.

That was not unusual in the Atlanta area, where there is a stable and flourishing Black middle class. We drove through somewhat homogeneous big-box houses, like Legos plunked down and spaced out. We passed Black men on golf carts and Black children shooting hoops in driveways. I marveled at the expanses of comfort, how in places like this Black achievement, affluence, and safety were normal rather than anomalous.

When we arrived at her house, she weaved a path among the bare studs, pointing to what was to be, beaming all the way. This wasn't my dream, but it was hers, and I was happy. I understood. For children who grew up in run-down houses in the middle of nowhere, having a new, modern house, even in the middle of nowhere, was a marker of "making it."

Although I always earned more money, it seemed to me that she had always lived more life, not with fancy things or extravagant vacations or starred restaurants. She reaped her

THE REUNION

joy in contentment and fulfillment. She reaped it from an environment that honored her.

I felt it every time I visited. My spirit was always set at ease. I always rested. I was the one of us who had run away up north and fought the world to make it see me. She was the one who had stayed down south where the world had folded her in its bosom and nurtured her.

It was the same feeling I got when visiting another college friend, Cassius, also a banker, just outside Washington, DC, in Prince George's County, home to some of the wealthiest Black communities in America.

It was these visits that helped draw me back south: moving from spaces where only concentrated poverty was most evident in Black communities, to ones where concentrated prosperity was also prevalent. And with this kind of economic opportunity, the cultural connectedness that it provided, the spiritual healing it promoted.

And it was these visits that suggested to me a higher calling, a moon shot ambition: to establish a region in the country in which white supremacy is not a prevailing ethos.

Beyond the push of state terror and oppressive, even lethal, neglect, and the pull of political power, purchasing power, and overall economic opportunity and possibility, there is, I believe, a spiritual, restorative need for the collective Black family to reunite.

Among America's descendants of slavery, there isn't one Black America, but two: the children of the Great Migration and the children of those who stayed behind in the South. I am a child of those who stayed, but I have had the great fortune to live half of my life in the South and the other half out of it.

That experience has informed my view that it is time to reunite and reconcile those two factions—the energetic enlightenment of the urban North and West with the patient, purposeful pragmatism of the South—because the amalgamation, somewhere in the middle, where the best of both melds, is where our greatest strength lives and can be manifest.

The Black narrative in America—constructed by America—is one of slavery and oppression, despair and deprivation, survival and overcoming. Anti-Black white supremacy is so omnipresent in the Black American story that it puts far out of reach what white America has been allowed to enjoy: the construction of a narrative of heroism, adventure, dominance, and invincibility.

They wrote the histories, the books, and the screenplays. The white man was a constant defender of righteousness and virtue, he was brave beyond comprehension, he was brilliant beyond belief. He could be plunked down in any foreign environment and conquer it: the Pilgrim in the New World, the cowboy on the western frontier, Tarzan in the jungles of Africa, an avatar on a foreign planet who rides the big red bird. The native, always darker people in his way were the lesser, pagans and savages, more emotional and less rational, deserving of their fate, destined to play the foil in the white man's fantasy.

Put another way, in the real world as well as the imagined one, white people in America have been acculturated to possession of power and possibility, and Black people to the absence of it.

THE REUNION

We need consolidated state power for the reclamation of memory and the faithful telling of history, from the initial telling of the tales, to the ability to influence or even control the textbooks from which our children are taught, textbooks that still do not fully incorporate Black history, Black humanity, and Black achievement.

We need a bloc of states—a region—in which we, and our children, are equally conditioned to success, support, and safety. We need a space in which Black imagination is equally encouraged, where we recognize that Black children dream too, that they gaze upon the same stars as all others, that adventure and invention are universally human traits that demand to be nurtured in all.

We need a space in which Black narrative can exist—and must exist—outside its relationship to anti-Black white supremacy. The Black story must be much more than slavery, oppression, and poverty. It must be so much bigger than the part of our literature white America chooses to reward: stories wrapped around racism, stories of struggling for self-acceptance and societal acceptance in a white world riven with that racism. We must have a space in which we can break free from the stories in cinema that white America chooses to celebrate—the junkies, the lawless, the brutal thugs, the abusive mother.

We need a space in which Black beauty is equally honored and exalted—hair that defies gravity rather than submits to it, noses wide at the base and undefined on the ridge, skin that absorbs the light rather than reflects it, melanin so abundant that it yellows the white of the eyes.

We need a space to remember that our trauma history is

not our total history. We have to break the chains of a violence we learn from the time we are children. Black people for generations, since enslavement, absorbed the violence white people visited upon them. They came to see pain as an expression of love, disfigurement as an expression of devotion. Black bodies had to be controlled, for their own survival. If a Black child couldn't come to control his body through practice, it would be controlled by pain. Better at the hands of the Black people who loved you than the white people who didn't.

We need space to reverse the absorption of white anxiety into our flesh—their fear of us, contempt for us, disdain of us. Black people in America have suffered so much oppression and deprivation that people—including many Black people—have falsely come to regard the resulting ill effect as a product *of* culture rather than an infliction *on* culture.

We taught our children that there was a different way of engaging the world and navigating it. For white children, exploration was encouraged. There was a sense of ownership of the space that they occupied. We, on the other hand, sat church-pew still. Obeisance was the objective. Dressing up and not acting out. Keeping quiet. Offering smiles easily, and controlling anger constantly. Playing against type to neutralize and expose the lie white people tell about you: that you are savage and unsophisticated, incapable of refinement or enlightenment.

We need an exodus to the South in sufficient numbers and density that Black people can come to know what real, lasting

THE REUNION

power feels like. Ultimately I believe this will be good not only for Black people, but for the country as a whole.

Black people have lived without autonomy for so long in this country that not only don't they know what the possession of it feels like, they in some ways harbor a subconscious fear of possessing it. The forbidden grew foreign, and frightful.

But there have been times when I have been able to glimpse Black people standing in the presence of their power, embracing it, and I have been able to imagine what a liberated people would look like.

One of those glimpses came in 2015 when I, along with a small group of Black journalists, was invited by the White House of the first Black president to travel with him on Air Force One to commemorate the fiftieth anniversary of "Bloody Sunday" in Selma, Alabama.

There were five of us: DeWayne Wickham, former *USA Today* columnist and founding dean of the School of Global Journalism and Communication at Morgan State University; April D. Ryan of American Urban Radio Networks; Rembert Browne of *Grantland*; and Zerlina Maxwell, a freelance journalist at the time.

Locals, brimming with pride and awe, waving and smiling, lined the roads that snaked through the woodlands, elated just to catch a glimpse of the motorcade. We waved back, although it wasn't clear to me if our windows were tinted and if they registered our reciprocity.

I had traveled to the South many times over the years since I moved up north. But this time was different. This time the North came with me; this time, the first Black president came with me. This time the nation came with me.

THE DEVIL YOU KNOW

When our van crested the Edmund Pettus Bridge, tens of thousands of people waiting at its foot came into view. April Ryan let out a scream that startled and amused us. The hair on my arms stood up. The historical gravity of the moment was made real, and I could see in it what Black power and Black history could be when it converged on a particular place. There were politicians and cultural leaders, activists from previous struggles and current ones, society notables and everyday people. Black people. Remembering, drawing strength from struggle, celebrating victory, preparing for the next wave of the war. All in the South, on southern soil, where most of our ancestors lived out their days. I was reminded once again that I needed to come home, and that I needed to bring the Black North with me.

There is so much Black blood soaked into the soil of the South that it belongs to us as much as anyone. As Jemar Tisby so insightfully wrote in *Vox*:

> *For black people, the South is our homeland away from home. We were divorced from our native soil on the African continent and shipped to agricultural regions of North America; the Deep South is as close as many African Americans will get to their past.*

There is something about the South that, as a Black person, speaks to me, spiritually. It is sounds that the air remembers. An enslaved woman's whisper and an enslaved man's wail. The crack of whip and flaying of flesh. The dogs. The dirges. But it is also the sounds of joy. The blues and the jazz. The Negro spiritual. Hambone slapping. Fraternities and sororities stepping. The boasts and belly laughs at family reunions.

This place, this earth, has the memory of our ancestors soaked into it. Not a tree grows that's not watered in some way by Black folks' tears and Black folks' blood. Some trees are old enough to remember the writhing that vibrated a branch like a caught fish on the line of a pole or the snap of a noosed neck that caused the movement to cease. Others, the respite of folks from the fields taking a break from the heat, playing the dozens and swigging sweet tea.

But today, nearly half of the Black people who once called the South home have been scattered to the other three corners of the country.

I say to Black Americans in destination cities: If you're happy and prospering in those cities, by all means stay. If you feel physically safe, economically secure, culturally celebrated, and spiritually edified, you have found your home.

But if that is not your lived experience, if you feel that the place you moved to or where you were born and raised has grown inhospitable, if you have tired of fighting the same battle that your parents fought, there is another option that is not only viable but desirable. I am saying that the Black South beckons.

I as much as anyone am drawn to the crackle of northern and western cities, to the pace, energy, and metabolism of these places. When I first arrived in New York in the early nineties, as an intern for the *New York Times*, I couldn't shake the feeling that I had hitherto existed in some sort of semi-slumber, and that being in the city was what being awake and alive were supposed to feel like, buildings tickling the sky, trains snaking underfoot, a blur of ambition and ingenuity, beauty and horror, an assault on the senses. There was a seductive muscularity to the city, an exquisite fierceness, a feeling of riding the razor between your destiny and your demise.

When I survey the New York City skyline now, I think it magnificent, an engineering marvel, a grand human achievement, but I'm also acutely aware that almost nothing I see was designed or financed by Black people. And I see a legacy of displacement and erasure. There are the buildings in the financial district built atop burial grounds of the enslaved. There was the eighteenth- and nineteenth-century free Black settlement called "Little Africa" in Greenwich Village whose residents shifted to housing uptown as new immigrants from Italy were arriving. There was the prosperous free Black community of Seneca Village destroyed to build Central Park. There was the Tenderloin community stretching for blocks around Herald Square, from which Black people were forced after a race riot. There was the Black and Puerto Rican neighborhood of San Juan Hill that was razed for the erecting of Lincoln Center.

Black people didn't simply land in Harlem because they saw it as the most desirable area in which to live. They were chased up the island of Manhattan to Harlem.

I moved to the North because of the pace of life and economic opportunities available to me there, but I soon found that my love of the North was not reciprocal. There is a clear hostility, in some ways a growing one, to communities of color in New York and other destination cities. My body had never stiffened when I walked past police officers until

I moved north. I had never felt uncomfortable in my own neighborhood until it fully gentrified, and most of the Black and Hispanic people moved out, and the new arrivals started to look at me like an interloper or a holdover.

Destination cities are the center of commerce, art, publishing (new media and old), television, movies, fashion, and advertising. They are homes to some of our greatest museums and greatest libraries. But Black people have to ask themselves whether they are truly beneficiaries of this bounty if they cannot fully access it, either because of low economic opportunity, infrastructural impediments, or the perception that these white spaces are unwelcoming to Black faces. And is the culture that Black people create in these cities honored and elevated, or is it treated as exoticism, novelty, spice?

In the South, Black culture is primary: it is expressed in powerful Black churches and impressive Black colleges, which, according to a *Newsweek* report citing United Negro College Fund data, "produce 70 percent of all Black dentists and doctors, 50 percent of Black engineers and public school teachers, and 35 percent of Black lawyers." Black culture is in the jazz and blues that were born in the South, and in the Black social welfare organizations and Black fellowship organizations that thrive there. It's in the nation's largest Black cultural festival, the Essence Festival, and the nation's largest Black sporting event, the Bayou Classic, both of which take place in New Orleans. It's in southern food, which the food historian Michael Twitty points out is really food of the enslaved, brought with them from the homeland and cooked by them here in America.

The South is the origin point of Black culture in America.

THE DEVIL YOU KNOW

It is impossible to be a proud Black person in America and to be ashamed of the South, to be averse to the South, to consider southern Black culture inferior and wanting. Southern heritage is intertwined with the very concept of American blackness. As Ossie Davis put it in his 1961 play *Purlie Victorious* about a Black preacher who returns to Jim Crow–era Georgia to fight to save a church, liberate Black people from the cotton field, and collect family restitution from a plantation owner:

> *I find, in being black, a thing of beauty: a joy; a strength; secret cup of gladness; a native land in neither time nor place—a native land in every Negro face! Be loyal to yourselves: your skin; your hair; your lips, your southern speech, your laughing kindness—are Negro kingdoms, vast as any other! Accept in full the sweetness of your blackness—not wishing to be red, nor white, nor yellow: nor any other race, or face, but this.*

Don't be ashamed of it, exalt it. Exalt what Maya Angelou called "the sweet language," adopted from a West African phrase, the way Black people sing the ends of sentences to connote endearment. Exalt the too-loud laugh that covers with cackling centuries of crying. Exalt it all. Exalt the Black Self. As the Texas-born Beyoncé, arguably the biggest pop star in America, daughter of an Alabama father and a Louisiana mother, sang, on "Black Parade," she was going back to the South, "where my roots ain't watered down / Growin', growin' like a baobab tree."

I am imploring Black people to be loyal to ourselves, to understand that we have in front of us an opportunity for

THE REUNION

self-determination of which our ancestors only dreamed, and about which white America has dreaded. This era's Hajj is the pull of Black people back to the piney hills of the Carolinas, Georgia, and Alabama, to the lowlands of Mississippi and the bayous of Louisiana.

The Great Migration was an extraordinary social experiment, a somewhat organic galvanizing of collective conscious that produced collective action. It unlocked opportunity and unleashed creativity for multiple generations of Black Americans. Its initial benefits have long since reached a point of diminishing returns, however; the successes of the Great Migration now stand shoulder to shoulder with the suffering that grew out of it. But there is a way to alter this reality.

The only thing Black people have to do is come home.

The South now beckons as the North once did. The promise of real power is made manifest. Seize it. Migrate. Move.

NOTES

Introduction

2 **6 percent of American adults:** Amanda Barroso and Rachel Minkin, "Recent Protest Attendees Are More Racially and Ethnically Diverse, Younger than Americans Overall," Pew Research Center, June 24, 2020, https://www.pewresearch.org/fact-tank/2020/06/24/recent-protest-attendees-are-more-racially-and-ethnically-diverse-younger-than-americans-overall/.

2 **racial reckoning:** Ailsa Chang, Rachel Martin, and Eric Marrapodi, "Summer of Racial Reckoning," National Public Radio, August 16, 2020, https://www.npr.org/2020/08/16/902179773/summer-of-racial-reckoning-the-match-lit.

2 **bottles of syrup:** Tiffany Hsu, "Aunt Jemima Brand to Change Name and Image Over 'Racial Stereotype,'" *New York Times*, June 17, 2020, https://www.nytimes.com/2020/06/17/business/media/aunt-jemima-racial-stereotype.html.

2 **bags of rice:** Jordan Valinsky, "Uncle Ben's and Mrs. Butterworth's Follow Aunt Jemima Phasing Out Racial Stereotypes in Logos," CNN, June 17, 2020, https://www.cnn.com/2020/06/17/business/uncle-bens-rice-racist/index.html.

3 **approval of Black Lives Matter protests:** Charles Franklin, "Black Lives Matter Protests in Wisconsin," Marquette Law School Poll, August 26, 2020, https://rpubs.com/PollsAndVotes/652966.

4 **"Vandalizing government buildings":** E. D. Mondainé, "Portland's Protests Were Supposed to Be About Black Lives. Now, They're White Spectacle," *Washington Post*, July 23, 2020, https://www.washingtonpost.com/opinions/2020/07/23/portlands-protests-were-supposed-be-about-black-lives-now-theyre-white-spectacle/.

5 **encouraged police brutality:** Melanie Eversley, "Trump Tells Law Enforcement: 'Don't Be Too Nice' with Suspects," *USA Today*, July 28, 2017, https://www.usatoday.com/story/news/2017/07/29/trump-tells-law-enforcement-dont-too-nice-suspects/522220001/.

NOTES

5 defended Confederate Monuments: Eugene Scott, "Trump's Ardent Defense of Confederate Monuments Continues as Americans Swing the Opposite Direction," *Washington Post*, July 1, 2020, https://www.washingtonpost.com/politics/2020/07/01/trumps-ardent-defense-confederate-monuments-continues-americans-swing-opposite-direction/.

6 majority of the redistricting: José Sepulveda, "Republicans Solidify Grip on State Legislatures, Which Is Likely to Lead to Redistricting and Gerrymandering Efforts in 2021," CNBC, November 6, 2020, https://www.cnbc.com/2020/11/06/republicans-to-lead-redistricting-efforts-in-many-states-in-2021.html.

6 exit polls conducted: 2020 presidential exit polls conducted by Edison Research for the National Election Pool, *New York Times*, https://www.nytimes.com/interactive/2020/11/03/us/elections/exit-polls-president.html.

6 share of the nonwhite vote: Zachary Evans, "Trump Won One-Quarter of Non-White Voters, Improving on 2016 Numbers: Exit Poll," *National Review*, November 4, 2020, https://www.nationalreview.com/news/trump-won-highest-share-of-non-white-vote-of-any-republican-since-1960-exit-polls-show/.

7 Associated Press's VoteCast: Estimates from A.P. VoteCast, a survey conducted for the Associated Press by NORC at the University of Chicago, *New York Times*, https://www.nytimes.com/interactive/2020/11/03/us/elections/ap-polls-georgia.html.

7 a quarter of the state's: 1990 Census of Population: General Population Characteristics, Georgia, https://www2.census.gov/library/publications/decennial/1990/cp-1/cp-1-12.pdf.

7 about a third of it: United States Census Bureau, Quick Facts, Georgia, https://www.census.gov/quickfacts/GA.

7 "251,000 Black people": Alana Semuels, "Reverse Migration Might Turn Georgia Blue," *The Atlantic*, May 23, 2018, https://www.theatlantic.com/politics/archive/2018/05/reverse-migration-might-turn-georgia-blue/560996/.

7 Abrams told NPR: Transcript of *All Things Considered* interview between Mary Louise Kelly and Stacey Abrams, National Public Radio, November 2, 2020, https://www.npr.org/2020/11/02/930504055/former-georgia-gubernatorial-candidate-on-a-push-for-voter-turnout.

7 Biden carried the state: "Georgia Confirms Biden Victory and Finds No Widespread Fraud after Statewide Audit," CNN, Friday, November

NOTES

20, 2020, https://www.cnn.com/2020/11/19/politics/georgia-recount-election-results/index.htm

1. The Past as Prologue

13 a grand parkway: Charles E. Beveridge, "Frederick Law Olmsted Sr., Landscape Architect, Author, Conservationist (1822–1903)," National Association for Olmsted Parks, http://www.olmsted.org/the-olmsted-legacy/frederick-law-olmsted-sr.

14 "People would trust": Mike Thomas, "Timuel Black," *Chicago*, January 16, 2019, http://www.chicagomag.com/Chicago-Magazine/February-2019/Timuel-Black/.

15 farther south: Joseph A. Hill, "The Recent Northward Migration of the Negro," *Opportunity* magazine, 1924.

15 only major exodus: Todd Arrington, "Exodusters," National Park Service, updated April 10, 2015, https://www.nps.gov/home/learn/historyculture/exodusters.htm.

15 "the best locality for the Negro": Frederick Douglass, *Life and Times of Frederick Douglass*, 1895, 531.

16 "the outstanding fact": R. H. Leavell, T. R. Snavely, T. J. Woofter, Jr., W. T. B. Williams, Francis D. Tyson, *Negro Migration in 1916–17* (Washington, DC: Government Printing Office, 1919), https://fraser.stlouisfed.org/files/docs/publications/dne/dne_migration1916-1917.pdf.

17 The Black population of Chicago: "DuSable to Obama," WTTW, https://interactive.wttw.com/dusable-to-obama/the-great-migration.

17 Some migrants traveled by bus: Chicago Historical Society, "Illinois Central Railroad Links to Chicago," *Encyclopedia of Chicago*, 2005, http://www.encyclopedia.chicagohistory.org/pages/3715.html.

17 "chicken bone express": Robert Norman Brown II, "Coming Home: Black Return Migration to the Yazoo-Mississippi Delta," Louisiana State University, LSU Digital Commons, 2001, https://digitalcommons.lsu.edu/cgi/viewcontent.cgi?article=1265&context=gradschool_disstheses.

17 Part of the mission: "National Urban League (NUL)," National Park Service, last updated April 1, 2016, https://www.nps.gov/articles/nationalurbanleague.htm.

17 "Do not loaf": "Chicago's Urban League Offers Assistance to Southern Migrants," American Social History Project, Center for Media and Learning, https://herb.ashp.cuny.edu/items/show/1597.

NOTES

18 "Should the influx": Kelly Miller, "Negro Migration: New Problems Raised by the Large Numbers Coming North," *New York Times*, September 9, 1916, https://timesmachine.nytimes.com/timesmachine/1916/09/09/104690354.pdf.

19 dubbed it the "Red Summer": Matthew Wills, "The Mob Violence of the Red Summer," *JSTOR Daily*, May 14, 2019, https://daily.jstor.org/the-mob-violence-of-the-red-summer/.

19 If you expand the definition of Red Summer: Associated Press, "Hundreds of Black Americans Were Killed During 'Red Summer.' A Century Later, Still Ignored," *USA Today*, July 23, 2019, https://www.usatoday.com/story/news/nation/2019/07/23/racial-violence-red-summer-1919-witnessed-white-black-murder/1802371001/.

19 "practically no new building": The Chicago Commission on Race Relations, *The Negro in Chicago: A Study of Race Relations and a Race Riot* (Chicago: University of Chicago Press, 1922), https://www.gutenberg.org/files/57343/57343-h/57343-h.htm.

19 "prevent lawlessness": Michael Jones-Correa, "American Riots: Structures, Institutions and History" (working paper, Russell Sage Foundation, 1999), https://studylib.net/doc/7508169/american-riots----russell-sage-foundation.

20 After the 1919 riots: Wendy Plotkin, "'Hemmed In': The Struggle against Racial Restrictive Covenants and Deed Restrictions in Post-WWII Chicago," *Journal of the Illinois State Historical Society* 94, no. 1 (Spring 2001): 39–69.

20 "To many black southerners": James R. Grossman, *Land of Hope: Chicago, Black Southerners, and the Great Migration* (Chicago: University of Chicago Press, 1989).

20 keep Black folks subordinate and separate: John R. Logan, Weiwei Zhang, and Miao Chunyu, "Emergent Ghettos: Black Neighborhoods in New York and Chicago, 1880–1940," *American Journal of Sociology* 120, no. 4 (January 2015): 1055–94.

21 "above all, there were two things": Lorraine Hansberry, adapted by Robert Nemiroff, *To Be Young, Gifted and Black: Lorraine Hansberry in Her Own Words* (New York: Vintage Books, 1995), 18.

22 "refugees and exiles of terror": Liz Mineo, "The Need to Talk About Race," *Harvard Gazette*, December 7, 2017, https://news.harvard.edu/gazette/story/2017/12/bryan-stevenson-seeks-national-conversation-about-slaverys-legacy/.

NOTES

22 Researchers believe that: David Eltis and Stanley L. Engerman, "Was the Slave Trade Dominated by Men?," *Journal of Interdisciplinary History* 23, no. 2 (Autumn 1992): 245, https://doi.org/10.2307/205275.

23 two and a half times the size of Africa's: Angus Maddison, *The World Economy* (Development Centre of the Organisation for Economic Co-operation and Development, 2006), Appendix B, 241, https://www.stat.berkeley.edu/~aldous/157/Papers/world_economy.pdf.

23 30 percent of that state's population: Campbell Gibson and Kay Jung, "Historical Census Statistics on Population Totals by Race, 1790 to 1990, and by Hispanic Origin, 1970 to 1990, for the United States, Regions, Divisions, and States," Table 55 (working paper no. 56, U.S. Census Bureau, 2002).

23 "migrating North increases": Dan A. Black, Seth G. Sanders, Evan J. Taylor, and Lowell J. Taylor, "The Impact of the Great Migration on Mortality of African Americans: Evidence from the Deep South," *American Economic Review* 105, no. 2 (February 2015): 477–503, https://pubs.aeaweb.org/doi/pdfplus/10.1257/aer.20120642.

24 "reproductive technology shock": George A. Akerlof and Janet L. Yellen, "New Mothers, Not Married: Technology Shock, the Demise of Shotgun Marriage, and the Increase in Out-of-Wedlock Births," Brookings Institution, 1996.

25 "were closely connected": Fengyan Tang, Heejung Jang, and Valire Carr Copeland, "Challenges and Resilience in African American Grandparents Raising Grandchildren: A Review of the Literature with Practice Implications," *GrandFamilies: The Contemporary Journal of Research, Practice and Policy* 2, no. 2 (September 2015), https://scholarworks.mich.edu/cgi/viewcontent.cgi?article=1018&context=grandfamilies.

25 "compared to a group": Morgan Sherburne, "The Economic Legacy of the Great Migration," *Michigan News*, University of Michigan, January 17, 2018, https://news.umich.edu/the-economic-legacy-of-the-great-migration/.

26 between 1877 and 1950: "Lynching in America: Confronting the Legacy of Racial Terror," Equal Justice Initiative, 2017.

26 nearly four out of five: Gregor Aisch, Robert Gebeloff, and Kevin Quealy, "Where We Came From and Where We Went, State by State," *New York Times*, August 19, 2014, https://www.nytimes.com/interactive/2014/08/13/upshot/where-people-in-each-state-were-born.html.

27 "America, after all, unscrambled": Julian Bond, Michael Long, eds., *Race Man: Selected Works, 1960–2015* (San Francisco: City Lights Books, 2020).

NOTES

2. The Proposition

31 three Southern states (South Carolina, Mississippi, and Louisiana) were majority Black: U.S. Census Bureau, "1870 Census: Volume 1. The Statistics of the Population of the United States," 1872, https://www2.census.gov/library/publications/decennial/1870/population/1870a-01.pdf.

32 a scant 8 percent: CNN Exit Polls, 2016, https://www.cnn.com/election/2016/results/exit-polls/south-carolina/senate.

32 only 6 percent: Ibid., https://www.cnn.com/election/2016/results/exit-polls/california/senate.

32 13.1 percent of those voting: "Electorate Profile: New Jersey," U.S. Census Bureau, 2014, https://www.census.gov/content/dam/Census/library/visualizations/2016/comm/cb16-tps60_graphic_voting_nj.jpg.

32 There have been only ten: "African American Senators," United States Senate, https://www.senate.gov/pagelayout/history/h_multi_sections_and_teasers/Photo_Exhibit_African_American_Senators.htm.

33 midwestern losses notwithstanding: CNN Presidential Results, 2016, https://www.cnn.com/election/2016/results/president.

34 "If a vocal minority": "President Nixon's 'Silent Majority' Speech on Vietnam War," C-SPAN, November 3, 1969, https://www.c-span.org/video/?153819-1/president-nixons-silent-majority-speech-vietnam-war.

35 "Overall, it was a mass gathering": Joan Herbers, "250,000 War Protesters Stage Peaceful Rally in Washington; Militants Stir Clashes Later," *New York Times*, November 15, 1969, https://archive.nytimes.com/www.nytimes.com/learning/general/onthisday/big/1115.html#article.

35 its first draft lottery: *Vietnam Magazine* staff, "What's Your Number? The Vietnam War Selective Service Lottery," Historynet.com, https://www.historynet.com/whats-your-number.htm.

36 "25 bombings—including": "Weather Underground Bombings," Federal Bureau of Investigation, https://www.fbi.gov/history/famous-cases/weather-underground-bombings.

36 Nixon stood before a map: "Nixon Announces Invasion of Cambodia," Roy Rosenzweig Center for History and New Media, George Mason University, http://chnm.gmu.edu/hardhats/cambodia.html#vietnamtime.

NOTES

36 "Taking Over Vermont": Richard Pollak, "Taking Over Vermont," *Playboy*, April 1972.

37 an obscure 1971 paper: James F. Blumstein and James Phelan, "Jamestown Seventy," *Yale Review of Law and Social Action* 1, no. 1 (1971), https://digitalcommons.law.yale.edu/cgi/viewcontent.cgi?article =1006&context=yrlsa.

38 the ground for a takeover had been seeded well: Yvonne Daley, *Going Up the Country: When the Hippies, Dreamers, Freaks, and Radicals Moved to Vermont* (Lebanon, NH: University Press of New England, 2018).

38 "a generation of hippies": Ibid.

38 5.8 percent of the population identifying as nonwhite: U.S. Census Bureau, 2019 estimates, https://www.census.gov/quickfacts /fact/table/VT/RHI125219#RHI125219.

38 a meager 0.4 percent: Campbell Gibson and Kay Jung, "Historical Census Statistics on Population Totals by Race, 1790 to 1990, and by Hispanic Origin, 1970 to 1990, for the United States, Regions, Divisions, and States" (working paper no. 56, U.S. Census Bureau, 2002), https://www.census.gov/content/dam/Census/library/working-papers /2002/demo/POP-twps0056.pdf.

39 Abbott came to call it the "Second Emancipation": Ethan Michaeli, "Bound for the Promised Land," *The Atlantic*, January 11, 2016, https://www.theatlantic.com/politics/archive/2016/01/chicago -defender/422583/.

40 The Road Ahead for Civil Rights: Courting Change: "Looking Back to Move Forward: Harry Belafonte and Dolores Huerta," Ford Foundation, July 17, 2013, https://www.fordfoundation.org/about /library/multimedia/courting-change-lunch-conversation-looking -back-to-move-forward/.

43 "was every much": John Nova Lomax, "Is Texas Southern, Western, or Truly a Lone Star?," *Texas Monthly*, March 3, 2015, https:// www.texasmonthly.com/the-daily-post/is-texas-southern-western-or -truly-a-lone-star/.

43 "some mix of New York": Kyle Munzenrieder, "19 Maps That Prove South Florida Is Not Really the South," *Miami New Times*, October 15, 2015, https://www.miaminewtimes.com/news/19-maps -that-prove-south-florida-is-not-really-the-south-7973020.

NOTES

43 the United States Census Bureau's designation: "Census Regions and Divisions in the United States," U.S. Census Bureau, https://www2.census.gov/geo/pdfs/maps-data/maps/reference/us_regdiv.pdf.

44 Only 6 percent: Walt Hickey, "Which States Are in the South?," FiveThirtyEight.com, April 30, 2014, https://fivethirtyeight.com/features/which-states-are-in-the-south/.

45 "How people have been": American Anthropological Association Statement on Race, May 17, 1998, https://www.americananthro.org/ConnectWithAAA/Content.aspx?ItemNumber=2583.

49 "The racist, misogynistic sentiment": Jesmyn Ward, "My True South: Why I Decided to Return Home," *Time*, July 26, 2018, https://time.com/5349517/jesmyn-ward-my-true-south/.

51 In our daily lives: Allen Gannett, "Here's How Your Brain Can Learn to Be Less Racist," *Fast Company*, June 7, 2018, https://www.fastcompany.com/40574302/heres-how-your-brain-can-learn-to-be-less-racist.

52 people want a specific kind of diverse neighborhood: Alvin Chang, "White America Is Quietly Self-Segregating," *Vox*, July 31, 2018, https://www.vox.com/2017/1/18/14296126/white-segregated-suburb-neighborhood-cartoon.

52 "In these highly segregated cities": Michael B. Sauter, Evan Comen, and Samuel Stebbins, "16 Most Segregated Cities in America," *24/7 Wall St.*, 2017, https://247wallst.com/special-report/2017/07/21/16-most-segregated-cities-in-america/.

53 an amendment to school integration: Matthew F. Delmont, *Why Busing Failed: Race, Media, and the National Resistance to School Desegregation* (Oakland: University of California Press, 2016), 98–99.

53 "The North is guilty": United Press International, "Excerpts from Ribicoff Rights Speech," *New York Times*, February 10, 1970, https://www.nytimes.com/1970/02/10/archives/excerpts-from-ribicoff-rights-speech.html.

53 "have hypocrisy in our hearts": Warren Weaver Jr., "Ribicoff Attacks Schools in North; Supports Stennis," *New York Times*, February 10, 1970, https://www.nytimes.com/1970/02/10/archives/ribicoff-attacks-schools-in-north-supports-stennis-backs-senate.html.

55 state revenue from Washington: Anne Stauffer, Justin Theal, and Brakeyshia Samms, "Federal Funds Hover at a Third of State Revenue," Pew Charitable Trusts, October 8, 2019, https://www.pewtrusts.org/en

NOTES

/research-and-analysis/articles/2019/10/08/federal-funds-hover-at-a-third-of-state-revenue.

55 "aside from Democrats": Lydia Saad, "U.S. Still Leans Conservative but Liberals Keep Recent Gains," Gallup, January 8, 2019, https://news.gallup.com/poll/245813/leans-conservative-liberals-keep-recent-gains.aspx.

55 like same-sex marriage: "Support for Same-Sex Marriage Grows, Even among Groups That Had Been Skeptical," Pew Research Center, June 26, 2017, https://www.pewresearch.org/politics/2017/06/26/support-for-same-sex-marriage-grows-even-among-groups-that-had-been-skeptical/. The survey found that 51 percent of Blacks support gay marriage compared to 47 percent of Republicans and 76 percent of Democrats.

55 the views of Black people: Frank Newport, "Blacks as Conservative as Republicans on Some Moral Issues," Gallup, December 3, 2008, https://news.gallup.com/poll/112807/blacks-conservative-republicans-some-moral-issues.aspx.

55 Black millennials are significantly more religious: Jeff Diamant and Besheer Mohamed, "Black Millennials Are More Religious Than Other Millennials," Pew Research Center, July 20, 2018, https://www.pewresearch.org/fact-tank/2018/07/20/black-millennials-are-more-religious-than-other-millennials/.

55 "If you'd walked into a gathering of older black folks": Karen Grigsby Bates, "Why Did Black Voters Flee the Republican Party in the 1960s?," National Public Radio, July 14, 2014, https://www.npr.org/sections/codeswitch/2014/07/14/331298996/why-did-black-voters-flee-the-republican-party-in-the-1960s.

56 "This is the first time": Harvard Sitkoff, *Toward Freedom Land: The Long Struggle for Racial Equality in America* (Lexington: University Press of Kentucky, 2010), 32.

57 "Goldwater can be seen": Karen Grigsby Bates, "Why Did Black Voters Flee the Republican Party in the 1960s?," National Public Radio, July 14, 2014, https://www.npr.org/sections/codeswitch/2014/07/14/331298996/why-did-black-voters-flee-the-republican-party-in-the-1960s.

57 just 13 percent of the Black vote: Theodore R. Johnson, "What Nixon Can Teach the GOP about Courting Black Voters," *Politico*, 2015, https://www.politico.com/magazine/story/2015/08/what-nixon-can-teach-the-gop-about-courting-black-voters-121392.

NOTES

57 "Negrophobe whites will quit": James Boyd, "Nixon's Southern Strategy 'It's All in the Charts,'" *New York Times Magazine*, May 17, 1970, https://www.nytimes.com/packages/html/books/phillips-southern.pdf.

58 "captured constituency": Paul Frymer, *Uneasy Alliances: Race and Party Competition in America* (Princeton, NJ: Princeton University Press, 1999), Apple Books e-book.

58 when people are forced: Farai Chideya, "Black Voters Are So Loyal That Their Issues Get Ignored," FiveThirtyEight.com, September 9, 2016, https://fivethirtyeight.com/features/black-voters-are-so-loyal-that-their-issues-get-ignored/.

59 "the Negro is the pawn": James Baldwin, *Notes of a Native Son*, (Boston: Beacon Press, 1955), Apple Books e-book.

61 "How much satisfaction": Zora Neale Hurston, "Letter to the *Orlando Sentinel*," *Orlando Sentinel*, August 11, 1955, https://teachingamericanhistory.org/library/document/letter-to-the-orlando-sentinel/.

63 "the very serious function of racism": Toni Morrison, "A Humanist View," a 1975 speech Morrison gave at Portland State University, https://www.mackenzian.com/wp-content/uploads/2014/07/Transcript_PortlandState_TMorrison.pdf.

3. The Push

70 Forty-one percent of Black people: Craig Palosky, "Poll: 7 in 10 Black Americans Say They Have Experienced Incidents of Discrimination or Police Mistreatment in Their Lifetime, Including Nearly Half Who Felt Their Lives Were in Danger," Kaiser Family Foundation, June 18, 2020, https://www.kff.org/disparities-policy/press-release/poll-7-in-10-black-americans-say-they-have-experienced-incidents-of-discrimination-or-police-mistreatment-in-lifetime-including-nearly-half-who-felt-lives-were-in-danger/.

72 As Yale explained: "Yale Police Release Internal Investigation Report," Yale University, March 3, 2015, https://news.yale.edu/2015/03/03/yale-police-release-internal-investigation-report.

73 He shot Rice: Charles M. Blow, "Tamir Rice and the Value of Life," *New York Times*, January 11, 2015, https://www.nytimes.com/2015/01/12/opinion/charles-m-blow-tamir-rice-and-the-value-of-life.html.

NOTES

75 January 2015 to August 2020: *Washington Post* database of police shootings, https://www.washingtonpost.com/graphics/investigations/police-shootings-database/.

76 "engage in a pattern": Violent Crime Control and Law Enforcement Act of 1994, Pub.L. 103–322, 108 Stat. 1796 (1994), https://www.congress.gov/103/bills/hr3355/BILLS-103hr3355enr.pdf.

76 thirteen police departments: "Police Reform and Accountability Accomplishments," Department of Justice, https://www.justice.gov/opa/file/797666/download.

76 These are cities: "An Interactive Guide to the Civil Rights Division's Police Reforms," Department of Justice, January 18, 2017, https://www.justice.gov/crt/page/file/922456/download.

78 "In the North, on the other hand": Martin Luther King Jr., "1. Next Stop: The North," *Saturday Review*, November 13, 1965.

80 "You bastards": William Bradford Huie, "The Shocking Story of Approved Killing in Mississippi," *Look* magazine, January 1956, https://www.pbs.org/wgbh/americanexperience/features/till-killers-confession/.

80 "I saw that his tongue was choked out": Interview with Mamie Till Mobley in the 2003 American Experience film, *American Experience*, "The Murder of Emmett Till," directed by Stanley Nelson, aired January 20, 2003, https://www.pbs.org/wgbh/americanexperience/films/till/#part01.

80 Tens of thousands: Devery S. Anderson, *Emmett Till: The Murder That Shocked the World and Propelled the Civil Rights Movement* (Jackson: University Press of Mississippi, 2015), Apple Books e-book.

80 one hundred thousand viewed: Craig Chamberlain, "60 Years Ago This Month, Emmett Till's Death Sparked a Movement," Illinois News Bureau, August 17, 2015, https://news.illinois.edu/view/6367/232501.

82 *Time* magazine would label this image: From *Time* magazine's collection of the hundred most influential photographs, as compiled by Ben Goldberger, Paul Moakley, and Kira Pollack: David Jackson, "Emmett Till," *Time*, 1955, http://100photos.time.com/photos/emmett-till-david-jackson.

82 hundreds of "disappeared" Black people: Margaret A. Burnham and Margaret M. Russell, "The Cold Cases of the Jim Crow

NOTES

Era," *New York Times*, August 28, 2015, https://www.nytimes.com /2015/08/28/opinion/the-cold-cases-of-the-jim-crow-era.html.

83 The beating was caught: Erik Ortiz, "George Holliday, Who Taped Rodney King Beating, Urges Others to Share Videos," NBC News, June 9, 2015, https://www.nbcnews.com/nightly-news/george -holliday-who-taped-rodney-king-beating-urges-others-share-n372551.

83 "officers taking turns swinging their nightsticks": Seth Mydans, "Tape of Beating by Police Revives Charges of Racism," *New York Times*, March 7, 1991, https://www.nytimes.com/1991/03/07/us /tape-of-beating-by-police-revives-charges-of-racism.html.

83 King's injuries included: "The Rodney King Affair:..." *Los Angeles Times*, March 24, 1991, http://articles.latimes.com/1991-03-24 /local/me-1422_1_king-s-injuries-officer-laurence-m-powell-beating.

83 "Gorillas in the Mist": Lois Timnick, "Judge Will Allow Race Evidence in King Case: Hearing: He Rules That Controversial Remarks Exchanged on Squad Car Computers May Be Introduced in the Trial of Four Officers Accused of Beating the Black Motorist," *Los Angeles Times*, June 11, 1991.

83 Those acquittals sparked: "The 10 Most-Costly Riots in the U.S.," *Chicago Tribune*, November 26, 2014, https://www.chicagotribune .com/chi-insurance-civil-unrest-riots-bix-gfx-20141126-htmlstory.html.

84 "the occupying army of a hostile government": Dennis Romero, "The Militarization of Police Started in Los Angeles," *LA Weekly*, August 15, 2014, https://www.laweekly.com/the-militarization -of-police-started-in-los-angeles/.

84 "We are surrounded": Monnica T. Williams, "The Link Between Racism and PTSD," *Psychology Today*, September 6, 2015, https://www.psychologytoday.com/us/blog/culturally-speaking /201509/the-link-between-racism-and-ptsd.

85 "That represents": Alyssa Fowers and William Wan, "Depression and Anxiety Spiked among Black Americans after George Floyd's Death," *Washington Post*, June 12, 2020, https://www.washingtonpost.com /health/2020/06/12/mental-health-george-floyd-census/.

86 "to help define for you the man I am running against": Andrew Rosenthal, "Bush and Dukakis Trade Accusations over Crime," *New York Times*, October 21, 1988, https://www.nytimes.com /1988/10/21/us/bush-and-dukakis-trade-accusations-over-crime.html.

NOTES

86 kept Byrne's badge in the Oval Office: Thomas Tracy, "President George H. W. Bush Formed Lasting Bond with the Family of a Slain NYPD Officer," *New York Daily News*, December 2, 2018, https://www.nydailynews.com/news/national/ny-metro-byrne-connection-george-bush-20181201-story.html.

86 "The racial disparities are staggering": Vanita Gupta, "The 40-Year War on Drugs: It's Not Fair, and It's Not Working," ACLU Washington, June 1, 2011, https://www.aclu-wa.org/blog/40-year-war-drugs-its-not-fair-and-its-not-working.

86 Black and Hispanic men: "Ending the War on Drugs: By the Numbers," Center for American Progress, June 27, 2018, https://www.americanprogress.org/issues/criminal-justice/reports/2018/06/27/452819/ending-war-drugs-numbers/.

87 "The pendulum protecting": David W. Dunlap, "System Lashed as Police Honor 5 Slain on Duty," *New York Times*, May 20, 1989, https://www.nytimes.com/1989/05/20/nyregion/system-lashed-as-police-honor-5-slain-on-duty.html.

87 "an ineffective and inefficient use of resources": David Muhlhausen, "How Congress Can Improve Its Financial Support for Law Enforcement," The Heritage Foundation, August 12, 2002, https://www.heritage.org/node/18791/print-display.

87 the program's financing: "Federal Spending Plan Slashes Anti-Crime Grants," Pew Charitable Trusts, December 31, 2007, https://www.pewtrusts.org/en/research-and-analysis/reports/2007/12/31/federal-spending-plan-slashes-anticrime-grants.

88 "Day and night": Press release from the National Association of Police Organizations, September 22, 2008, http://p2008.org/interestg08/napo092208pr.html.

89 "no better friends of law enforcement": Press release from Obama for America, July 25, 2012, http://www.p2012.org/interestg/napo072512.html.

89 "We have all seen": "Statement by Attorney General Holder on Federal Investigation into Death of Eric Garner," United States Department of Justice, December 3, 2014, https://www.justice.gov/opa/speech/statement-attorney-general-holder-federal-investigation-death-eric-garner.

89 When two New York City police: Benjamin Mueller and Al Baker,

NOTES

"2 N.Y.P.D. Officers Killed in Brooklyn Ambush; Suspect Commits Suicide," *New York Times*, December 20, 2014, https://www.nytimes.com/2014/12/21/nyregion/two-police-officers-shot-in-their-patrol-car-in-brooklyn.html.

90 "four months of propaganda": Laura Carroll, "Giuliani: Obama 'Propaganda' Says 'Everybody Should Hate the Police,'" Politifact, The Poynter Institute, December 23, 2014, https://www.politifact.com/factchecks/2014/dec/23/rudy-giuliani/giuliani-obama-propaganda-says-everybody-should-ha/.

90 "Politicians have created an environment": *The Washington Report*, The Newsletter of the National Association of Police Organizations, December 22, 2014, http://www.napo.org/files/9814/1927/8243/Washington_Report_-_December_22_2014.docx.pdf.

91 "all persons of color": "After Slavery: Educator Resources," Lowcountry Digital History Initiative, https://ldhi.library.cofc.edu/exhibits/show/after_slavery_educator/unit_three_documents/document_eight.

91 Freedmen without: United States Congressional serial set, Volume 1276, 212.

92 Roof was there: Tonya Maxwell, "Witness in Dylann Roof Trial: 'I Was Just Waiting My Turn,'" *Citizen Times*, December 7, 2016, https://www.citizen-times.com/story/news/local/2016/12/07/witness-dylann-roof-trial-just-waiting-my-turn/95082762/.

92 As *The Atlantic* summarized: Ethan Michaeli, "Bound for the Promised Land," *The Atlantic*, January 11, 2016, https://www.theatlantic.com/politics/archive/2016/01/chicago-defender/422583/.

94 caused the death mask: Teddy Kulmala and Bristow Marchant, "75 Years Ago Today, SC Executed a Black Teenager after a Three-Hour Trial," *The State*, June 16, 2019, https://www.thestate.com/news/local/crime/article231467578.html.

96 nearly 110,000 Black people: Brentin Mock, "Where Gentrification Is an Emergency, and Where It's Not," Bloomberg CityLab, April 5, 2019, https://www.citylab.com/equity/2019/04/where-gentrification-happens-neighborhood-crisis-research/586537/.

97 There are none in Mississippi: Data from the Southern Poverty Law Center's "Hate Map," 2019, https://www.splcenter.org/fighting-hate/extremist-files/ideology/white-nationalist.

NOTES

98 "the most concentrated cluster of racist searches": David H. Chae, Sean Clouston, Mark L. Hatzenbuehler, Michael R. Kramer, Hannah L. F. Cooper, Sacoby M. Wilson, Seth I. Stephens-Davidowitz, Robert S. Gold, and Bruce G. Link, "Association between an Internet-Based Measure of Area Racism and Black Mortality," *PLOS ONE* 10, no. 4 (2015), https://journals.plos.org/plosone/article?id=10.1371/journal.pone.0122963#pone.

98 released from active duty: "James Alex Fields Jr.: Charlottesville Suspect Was in the Army, Spokesperson Says," WCPO, August 13, 2017, https://www.wcpo.com/news/james-alex-fields-jr-charlottesville-suspect-was-in-the-army-spokesperson-says.

98 the Proud Boys: "Proud Boys" The Southern Poverty Law Center, https://www.splcenter.org/fighting-hate/extremist-files/group/proud-boys.

98 described as having gone: Alan Feuer, "Proud Boys Founder: How He Went from Brooklyn Hipster to Far-Right Provocateur," *New York Times Magazine*, October 16, 2018, https://www.nytimes.com/2018/10/16/nyregion/proud-boys-gavin-mcinnes.html.

99 In 2015 four out of five: Lauren Musu-Gillette, Anlan Zhang, Ke Wang, Jizhi Zhang, Jana Kemp, Melissa Diliberti, and Barbara A. Oudekerk, *Indicators of School Crime and Safety: 2017* (Washington, DC: National Center for Education Statistics, Institute of Education Sciences, 2018), https://nces.ed.gov/pubs2018/2018036.pdf.

99 white supremacists targeting: "White Supremacists Double Down on Propaganda in 2019," Anti-Defamation League, February 2020, https://www.adl.org/media/14038/download.

99 "the Creative Class": Carey L. Biron, "Washington D.C. Sued for Putting 'Creative Class' above Minority Communities," *Christian Science Monitor*, June 28, 2018, https://www.csmonitor.com/USA/Society/2018/0628/Washington-D.C.-sued-for-putting-creative-class-above-minority-communities.

100 "Chocolate Chip City": DeNeen L. Brown, "'The End of Our Journey': A Historic Black Church Closes Its Doors in a Changing D.C.," *Washington Post*, September 30, 2018, https://www.washingtonpost.com/local/the-end-of-our-journey-a-historic-black-church-closes-its-doors-in-a-changing-dc/2018/09/30/b2f3f222-c1c5-11e8-a1f0-a4051b6ad114_story.html.

NOTES

101 As Fagan wrote: Taken from trial documents in *David Floyd, Lalit Clarkson, Deon Dennis Ourlicht, on behalf of themselves and all others similarly situated, v. the City of New York, et al.*, argued before the United States Court of Appeals for the Second District, 2012, https://cases.justia.com/federal/appellate-courts/ca2/13-3088/359/0.pdf?ts=1394575625.

101 "Young black and Latino men": "Stop-and-Frisk 2011," New York Civil Liberties Union briefing, May 9, 2012, https://www.nyclu.org/sites/default/files/publications/NYCLU_2011_Stop-and-Frisk_Report.pdf.

102 young man named Tyquan: Julie Dressner and Edwin Martinez, "The Scars of Stop-and-Frisk," *New York Times*, June 12, 2012, https://www.nytimes.com/2012/06/12/opinion/the-scars-of-stop-and-frisk.html.

102 A 2010 investigation: Ray Rivera, Al Baker, and Janet Roberts, "A Few Blocks, 4 Years, 52,000 Police Stops," *New York Times*, July 12, 2010, https://www.nytimes.com/2010/07/12/nyregion/12frisk.html.

102 "That's true in virtually every city": Ibid.

102 at the height: Quinnipiac Poll of NYC residents, March 13, 2012, and June 14, 2012, https://poll.qu.edu/new-york-city/release-detail?ReleaseID=1716; https://poll.qu.edu/new-york-city/release-detail?ReleaseID=1764.

103 The majority of white people: Quinnipiac Poll of NYC residents, October 21, 2013, https://poll.qu.edu/new-york-city/release-detail?ReleaseID=1967.

103 "Eighty-seven percent": Charles M. Blow, "You Must Never Vote for Bloomberg," *New York Times*, November 10, 2019, https://www.nytimes.com/2019/11/10/opinion/michael-bloomberg.html.

104 In 2013, only: *Diversity on the Force: Where Police Don't Mirror Communities*, Governing Special Report, September 2015, https://images.centerdigitaled.com/documents/policediversityreport.pdf.

104 "use gun force": Mark Hoekstra and CarlyWill Sloan, "Does Race Matter for Police Use of Force? Evidence from 911 Calls," National Bureau of Economic Research, February 2020, http://conference.nber.org/conf_papers/f142725.pdf.

105 Wayne State University: Brad W. Smith, "The Impact of Police Officer Diversity on Police-Caused Homicides," *Policy Studies Journal* 31,

NOTES

no. 2 (May 2003): 147–62, https://onlinelibrary.wiley.com/doi/abs/10.1111/1541-0072.t01-1-00009.

106 A 2020 analysis by the Marshall Project: Eli Hager and Weihua Li, "A Major Obstacle to Police Reform: The Whiteness of Their Union Bosses," The Marshall Project, June 10, 2020, https://www.themarshallproject.org/2020/06/10/a-major-obstacle-to-police-reform-the-whiteness-of-their-union-bosses.

106 2017 Pew Research Center report: Rich Morin, Kim Parker, Renee Stepler, and Andrew Mercer, "Behind the Badge," Pew Research Center, January 11, 2017, https://www.pewsocialtrends.org/2017/01/11/behind-the-badge/.

107 New York City raised: "New York City Fine Revenues Update," Report from New York City Comptroller Scott M. Stringer, May 3, 2017, https://comptroller.nyc.gov/reports/new-york-city-fine-revenues-update/.

107 shifting the mission of entire police departments: Jack Hitt, "Police Shootings Won't Stop Unless We Also Stop Shaking Down Black People," *Mother Jones*, September/October 2015, https://www.motherjones.com/politics/2015/07/police-shootings-traffic-stops-excessive-fines/.

108 During Bloomberg's mayoralty: Brendan Cheney, "Racial Disparities Persist in New York City Marijuana Arrests," *Politico*, February 13, 2018, https://www.politico.com/states/new-york/city-hall/story/2018/02/13/racial-disparities-continue-in-new-york-city-marijuana-arrests-248896.

108 according to an analysis by: Thomas Kaplan, "Cuomo Seeks Cut in Frisk Arrests," *New York Times*, June 3, 2012, https://www.nytimes.com/2012/06/04/nyregion/cuomo-seeks-cut-in-stop-and-frisk-arrests.html.

108 research consistently finds: Fred Dews, "Charts of the Week: Marijuana Use by Race, Islamist Rule in Middle East, Climate Adaptation Savings," Brookings Institution, August 11, 2017, https://www.brookings.edu/blog/brookings-now/2017/08/11/charts-of-the-week-marijuana-use-by-race/.

108 "stopping and frisking more than a half million": Press release from the Drug Policy Alliance, "New Data Released: NYPD Made More Marijuana Possession Arrests in 2011 Than in 2010; Illegal Searches and Manufactured Misdemeanors Continue Despite Order

NOTES

by Commissioner Kelly to Halt Unlawful Arrests," January 31, 2012, https://www.drugpolicy.org/press-release/2012/01/new-data-released-nypd-made-more-marijuana-possession-arrests-2011-2010.

108   "One of the unintended": Bobby Allyn, "'Throw Them Against the Wall and Frisk Them': Bloomberg's 2015 Race Talk Stirs Debate," National Public Radio, February 11, 2020, https://www.npr.org/2020/02/11/804795405/throw-them-against-the-wall-and-frisk-them-bloomberg-s-2015-race-talk-stirs-deba.

108   New York City saw: "Fees, Fines and Fairness: How Monetary Charges Drive Inequality in New York City's Criminal Justice System," Report from New York City Comptroller Scott M. Stringer, September 10, 2019, https://comptroller.nyc.gov/reports/fees-fines-and-fairness/.

109   transferred $17.5 million: Ibid.

109   $53 million was posted in cash bail: "The Public Cost of Private Bail: A Proposal to Ban Bail Bonds in NYC," Report from New York City Comptroller Scott M. Stringer, January 17, 2018, https://comptroller.nyc.gov/reports/the-public-cost-of-private-bail-a-proposal-to-ban-bail-bonds-in-nyc/.

109   "levied $19,386,418,544 in money bail": Jeannette Chi, "The Price for Freedom: Bail in the City of L.A. A Million Dollar Hoods Report," UCLA Ralph J. Bunche Center for African American Studies, December 5, 2017, https://bunchecenter.ucla.edu/2017/12/05/the-price-for-freedom-bail-in-the-city-of-l-a-a-million-dollar-hoods-report/.

110   "Louisiana's incarceration rate is": Cindy Chang, "Louisiana Is the World's Prison Capital," *Times-Picayune*, May 13, 2012, https://www.nola.com/news/crime_police/article_8feef59a-1196-5988-9128-1e8e7c9aefda.html.

111   "from 2010 to 2014": Campbell Robertson, "The Prosecutor Who Says Louisiana Should 'Kill More People,'" *New York Times*, July 8, 2015, https://www.nytimes.com/2015/07/08/us/louisiana-prosecutor-becomes-blunt-spokesman-for-death-penalty.html.

112   in both 2008: *New York Times*' presidential election results for 2008, https://www.nytimes.com/elections/2008/results/president/votes.html.

112   and 2012: *New York Times*' presidential election results for 2012, https://www.nytimes.com/elections/2012/results/president/big-board.html.

NOTES

112 Vermont Public Radio poll: Bayla Metzger and Mitch Wertlieb, "'Do You Think Racism Is a Problem in Vermont Today?': Poll Looks At State's Perception," Vermont Public Radio, October 24, 2018, https://www.vpr.org/post/do-you-think-racism-problem-vermont-today-poll-looks-states-perception#stream/0.

112 resigned her post: Liam Stack, "Black Female Lawmaker in Vermont Resigns after Racial Harassment," *New York Times*, September 26, 2018, https://www.nytimes.com/2018/09/26/us/politics/kiah-morris-vermont.html.

112 "Racist, sexist, anti-gay": Katharine Q. Seelye, "Protesters Disrupt Speech by 'Bell Curve' Author at Vermont College," *New York Times*, March 3, 2017, https://www.nytimes.com/2017/03/03/us/middlebury-college-charles-murray-bell-curve-protest.html.

112 "As soon as he": Will Digravio, "Why Two Middlebury Football Players Decided to Kneel During the National Anthem," *Middlebury Campus*, December 6, 2017, https://middleburycampus.com/37226/news/inside-midd-football-players-decision-to-kneel-during-anthem/.

113 "Abolitionists launched the": Jane Williamson, "History Space: Radical Abolition in Vermont," *Burlington Free Press*, September 2, 2017, https://www.burlingtonfreepress.com/story/news/2017/09/02/history-space-radical-abolition-vermont/105202042.

113 Just 5 percent: "Census 2020 California Hard-to-Count Fact Sheet San Francisco County," US Census Bureau, 2020, https://census.ca.gov/wp-content/uploads/sites/4/2019/06/San-Francisco.pdf.

113 "The reality is": Speech tweeted by Chesa Boudin (@ChesaBoudin), Twitter, September 12, 2019, 2:28 a.m., https://twitter.com/chesaboudin/status/1172034132437655553?lang=en.

114 age-adjusted mortality rate: "The Color of Coronavirus: COVID-19 Deaths by Race and Ethnicity in the U.S.," APM Research Lab, August 18, 2020, https://www.apmresearchlab.org/covid/deaths-by-race.

114 the Black unemployment rate: Jhacova Williams, "Latest Data: Black-White and Hispanic-White Gaps Persist as States Record Historic Unemployment Rates in the Second Quarter," Economic Policy Institute, August 2020, https://www.epi.org/indicators/state-unemployment-race-ethnicity/.

115 "To be a Negro in this country": Joan Didion, *We Tell Ourselves Stories in Order to Live: Collected Nonfiction* (New York: Alfred A. Knopf, 2006), 199.

NOTES

4. The Pull

121 The panel has been: Panel at the All Black National Convention, Atlanta, Georgia, 2016, https://youtu.be/kM0wBiSGW-8.

122 Every city on the list: Samuel Stebbins and Evan Comen, "These Are the 15 Worst Cities for Black Americans," *USA Today*, November 16, 2018, https://www.usatoday.com/story/money/2018/11/16/racial-disparity-cities-worst-metro-areas-black-americans/38460961/.

123 many of those cities were in the South: Joel Kotkin, "The Cities Where African-Americans Are Doing the Best Economically," *Forbes*, January 15, 2018, https://www.forbes.com/sites/joelkotkin/2018/01/15/the-cities-where-african-americans-are-doing-the-best-economically-2018/#7b4e8dc21abe.

123 southern cities topped the list: Dean Stansel, "Ranking U.S. Metropolitan Areas on the Economic Freedom Index," Reason Foundation, January 31, 2019, https://reason.org/policy-study/us-metropolitan-area-economic-freedom-index/.

123 "Southeastern states have a higher concentration of Black businesses": Felecia Hatcher, "Blacktech Week's 2017 List of the Best Cities for Black Owned Businesses," Huffington Post, September 26, 2017, https://www.huffpost.com/entry/blacktech-weeks-2017-list-of-the-best-cities-for-black_b_59c9d86ae4b08d6615504575.

124 A 2019 Brookings report: Alan Berube, "Black Household Income Is Rising across the United States," Brookings Institution, October 3, 2019, https://www.brookings.edu/blog/the-avenue/20193/black-household-income-is-rising-across-the-united-states/.

124 "accounting for regional": David Autor, "The Faltering Escalator of Urban Opportunity," Research Brief, October 3, 2020, https://workofthefuture.mit.edu/wp-content/uploads/2020/09/2020-Research-Brief-Autor.pdf.

125 "vast vote colonization": "Watch for Vote Frauds," *New York Times*, October 18, 1916, https://timesmachine.nytimes.com/timesmachine/1916/10/18/301920542.pdf.

126 "ascertain the names": "Tracing Negro 'Floaters,'" *New York Times*, October 28, 1916, https://timesmachine.nytimes.com/timesmachine/1916/10/28/100225335.pdf.

126 Jackson's strategy was straightforward: Taken from the 2017

NOTES

movie *Maynard*, directed by Samuel D. Pollard, written by Wendy Eley Jackson and Samuel D. Pollard.

127 "the subjugated land": Nicholas D. Kimble, "Republic of New Afrika," Brown-Tougaloo Cooperative Exchange, http://cds.library.brown.edu/projects/FreedomNow/themes/blkpower/.

127 the RNA's Declaration of Independence: New Afrikan People's Organization, "New Afrikan Declaration of Independence," Freedom Archives, 1968, https://www.freedomarchives.org/Documents/Finder/DOC513_scans/NAPO/513.NAPO.NewAfrikanDec.pdf.

128 82,000 Black millennials: Reniqua Allen, "Racism Is Everywhere, So Why Not Move South?," *New York Times*, July 8, 2017, https://www.nytimes.com/2017/07/08/opinion/sunday/racism-is-everywhere-so-why-not-move-south.html.

129 Among Black people: "Metro Atlanta Population Growth Fueled By Minorities," *Atlanta Journal-Constitution*, June 24, 2019, https://www.ajc.com/news/local/metro-atlanta-population-growth-fueled-minorities/fz4aXo7CdyhEai1RgjW8jO/.

131 Harlem's Black population: Colby Hamilton, "Can Brooklyn's Historical Black Congressional District Survive?," WNYC, December 5, 2011, https://www.wnyc.org/story/196003-can-brooklyns-historical-black-congressional-district-survive/.

131 1,200 majority-Black cities: Andre M. Perry, "Recognizing Majority-Black Cities, When Their Existence Is Being Questioned," Brookings Institution, October 5, 2017, https://www.brookings.edu/blog/the-avenue/2017/10/04/recognizing-majority-black-cities-when-their-existence-is-being-questioned/.

131 nearly a quarter: Ibid.

131 Sierra Leone: Central Intelligence Agency, *World Factbook*, s.v. "Sierra Leone," https://www.cia.gov/library/publications/the-world-factbook/geos/print_sl.html.

131 Sierra Leone was a major source: James Knight and Katrina Manson, "Sierra Leone Draws Americans Seeking Slave Roots," Reuters, March 22, 2007. Analysis by professor Joseph Opala of James Madison University.

132 A 2017 Pew Research Center report: Monica Anderson, "African Immigrant Population in the U.S. Steadily Climbs," Pew Research Center, February 14, 2017, http://www.pewresearch.org/fact-tank/2017/02/14/african-immigrant-population-in-u-s-steadily-climbs/.

NOTES

132 forty-two when elected: Benjamin Freed, "Muriel Bowser Elected DC Mayor," *Washingtonian*, November 4, 2014, https://www.washingtonian.com/2014/11/04/muriel-bowser-elected-dc-mayor/.

132 thirty-five when elected: Ned Oliver, "Levar Stoney, Richmond's Youngest Elected Mayor, Took Office Sunday: Here's How He Got There," *Richmond Times-Dispatch*, December 31, 2016, https://richmond.com/news/local/levar-stoney-richmonds-youngest-elected-mayor-took-office-sunday-heres-how-he-got-there/article_bc314e58-acdb-5c19-a522-d95c008fe546.html.

133 four states: Teresa Wiltz, "Talk of Reparations for Slavery Moves to State Capitols," Pew Charitable Trusts, October 3, 2019, https://www.pewtrusts.org/en/research-and-analysis/blogs/stateline/2019/10/03/talk-of-reparations-for-slavery-moves-to-state-capitols.

133 the reparations provision: Grace Toohey, "New Florida Law to Teach, Recognize 1920 Ocoee Massacre That Destroyed City's Black Community," *Orlando Sentinel*, June 24, 2020.

133 city council of Asheville: Shawna Mizelle, "North Carolina City Votes to Approve Reparations for Black Residents," CNN, July 15, 2020, https://www.cnn.com/2020/07/15/us/north-carolina-asheville-reparations/index.html.

134 Black people are heavy: Monica Anderson, Skye Toor, Lee Rainie, and Aaron Smith, "Activism in the Social Media Age," Pew Research Center, July 11, 2018, https://www.pewresearch.org/internet/2018/07/11/activism-in-the-social-media-age/.

134 Black people represented: Rani Molla, "How Facebook Compares to Other Tech Companies in Diversity," Recode, April 11, 2018, https://www.recode.net/2018/4/11/17225574/facebook-tech-diversity-women.

134 When *Adweek* published: Ivie Ani, "Adweek Reveals List of 100 CEOs, Media, Tech Leaders and There's Zero Black Executives on It," Okayplayer, http://www.okayplayer.com/news/adweek-100-ceos-power-list-no-black-executives.html.

134 The majority of players: Charles Mudede, "Why the Overrepresentation of Black Americans in Professional Sports Is Not a Good Thing," *The Stranger*, September 25, 2017, https://www.thestranger.com/slog/2017/09/25/25432524/why-the-over-representation-of-black-americans-in-professional-sports-is-not-a-good-thing.

NOTES

135 The NFL just, in 2020: Kevin Patra, "Washington Makes Jason Wright First Black Team President," National Football League, August 17, 2020, https://www.nfl.com/news/washington-taps-former-rb-jason-wright-to-be-team-president.

135 historically Black colleges: Mark Armour and Daniel R. Levitt, "Baseball Demographics, 1947–2016," Society for American Baseball Research, https://sabr.org/bioproj/topic/baseball-demographics-1947-2012.

135 "Bringing elite athletic talent": Jemele Hill, "It's Time for Black Athletes to Leave White Colleges," *The Atlantic*, October 2019, https://www.theatlantic.com/magazine/archive/2019/10/black-athletes-should-leave-white-colleges/596629/.

136 "It makes sense to give": Andrew Barker, "Sean Combs Slams Industry's Lack of Investment in Black Enterprise, Previews Next Moves," *Variety*, July 10, 2018, https://variety.com/2018/music/features/sean-combs-diddy-black-enterprise-tv-music-1202868715/.

136 $1.2 trillion in purchasing power: Ellen McGirt, "raceAhead: A New Nielsen Report Puts Black Buying Power at $1.2 Trillion," *Fortune*, February 28, 2018, https://fortune.com/2018/02/28/raceahead-nielsen-report-black-buying-power/.

138 state preemption of municipal power is on the rise: National League of Cities, "State Preemption of Local Authority Continues to Rise, According to New Data from the National League of Cities," press release, April 5, 2018, https://www.nlc.org/article/state-preemption-of-local-authority-continues-to-rise-according-to-new-data-from-the.

138 The vast majority: Wendy Sawyer and Peter Wagner, "Mass Incarceration: The Whole Pie 2020," Prison Policy Initiative, press release, March 24, 2020, https://www.prisonpolicy.org/reports/pie2020.html.

139 There is also a persistent: Dick Startz, "The Achievement Gap in Education: Racial Segregation versus Segregation by Poverty," Brookings Institution, January 20, 2020, https://www.brookings.edu/blog/brown-center-chalkboard/2020/01/20/the-achievement-gap-in-education-racial-segregation-versus-segregation-by-poverty/.

139 state and local levels: Michael Leachman and Eric Figueroa, "K-12 School Funding Up in Most 2018 Teacher-Protest States, But Still Well Below Decade Ago," Center on Budget and Policy Priorities,

NOTES

March 6, 2019, https://www.cbpp.org/research/state-budget-and-tax/k-12-school-funding-up-in-most-2018-teacher-protest-states-but-still.

139 "State governments exercise": "How Are the Local, State and Federal Governments Involved in Education? Is This Involvement Just?," The Center for Public Justice, https://www.cpjustice.org/public/page/content/cie_faq_levels_of_government.

139 "about 7 percent": Madeline Will, "65 Years after 'Brown v. Board,' Where Are All the Black Educators?," *Education Week*, May 14, 2019, https://www.edweek.org/ew/articles/2019/05/14/65-years-after-brown-v-board-where.html?r=1881250887.

139 allocating about $300 billion: "Health and Hospital Expenditures," Urban Institute. The report reads, "In 2017, state and local governments spent $294 billion, or 10 percent of direct general spending, on health and hospitals. Health and hospitals combined were the fourth-largest source of state and local direct general spending in 2017 and roughly equal to higher education expenditures."

139 "The link between": Robert J. Sampson and Alix S. Winter, "The Racial Ecology of Lead Poisoning: Toxic Inequality in Chicago Neighborhoods, 1995–2013," *Du Bois Review* 13, no. 2 (Fall 2016): 261, https://doi.org/10.1017/S1742058X16000151.

140 "The pandemic is blowing": John C. Austin, "COVID-19 Is Turning the Midwest's Long Legacy of Segregation Deadly," Brookings Institution, April 17, 2020, https://www.brookings.edu/blog/the-avenue/2020/04/17/covid-19-is-turning-the-midwests-long-legacy-of-segregation-deadly/.

140 "In the decades": "HIV in the Southern United States," Issue Brief from the US Centers for Disease Control, September 2019, https://www.cdc.gov/hiv/pdf/policies/cdc-hiv-in-the-south-issue-brief.pdf.

141 And nearly half: Ibid.

141 "single largest source": "Medicaid and HIV," The Henry J. Kaiser Family Foundation fact sheet, October 2016, http://files.kff.org/attachment/Fact-Sheet-Medicaid-and-HIV.

141 "fulfilling a campaign promise": Richard Fausset and Abby Goodnough, "Louisiana's New Governor Signs an Order to Expand Medicaid," *New York Times*, January 12, 2016, https://www.nytimes.com/2016/01/13/us/louisianas-new-governor-signs-an-order-to-expand-medicaid.html.

NOTES

141 That year Louisiana: "HIV Infection and AIDS, 2016," Minnesota Department of Health, 2016, https://www.health.state.mn.us/diseases/reportable/dcn/sum16/hiv.html.

141 dropped by 12 percent: Emily Woodruff, "New HIV Cases in Louisiana Hit Decade Low in 2018; Health Officials Hopeful for Epidemic's End," Nola.com, July 13, 2019, https://www.nola.com/news/healthcare_hospitals/article_4c7e9078-9dd0-11e9-8d81-cf4844533a8d.html.

5. The End of Hoping and Waiting

149 "I'm not sure if you've": Veronica Stracqualursi, "Obama Honored with RFK Human Rights Award," CNN, December 13, 2018, https://www.cnn.com/2018/12/13/politics/obama-rfk-ripple-of-hope-award/index.html.

151 "crafted over centuries": Andre C. Willis, "What Hope Actually Meant to Martin Luther King Jr.," Big Think, July 5, 2017, https://bigthink.com/videos/andre-willis-what-hope-actually-meant-to-martin-luther-king-jr.

153 "Each time a man": Robert F. Kennedy, "Day of Affirmation Address," University of Capetown, Capetown, South Africa, June 6, 1966, https://www.jfklibrary.org/learn/about-jfk/the-kennedy-family/robert-f-kennedy/robert-f-kennedy-speeches/day-of-affirmation-address-university-of-capetown-capetown-south-africa-june-6-1966.

153 "Think about how many": Barack Obama, "Remarks by President Obama at the University of Cape Town," South Africa, June 30, 2013, transcript, The White House, https://obamawhitehouse.archives.gov/the-press-office/2013/06/30/remarks-president-obama-university-cape-town.

155 The tests: Charles M. Blow, "A Nation of Cowards?," *New York Times*, February 20, 2009, https://www.nytimes.com/2009/02/21/opinion/21blow.html.

155 according to exit polls: 2016 exit polls as published by CNN, https://www.cnn.com/election/2016/results/exit-polls.

156 "according to exit polls": 2018 exit polls as published by CNN, https://www.cnn.com/election/2018/exit-polls.

156 In July 2020: "Biden Widens Lead over Trump to 15 Points in Presidential Race, Quinnipiac University National Poll Finds; Trump Job Approval Rating Drops to 36 Percent," Quinnipiac University Poll,

NOTES

July 15, 2020, https://poll.qu.edu/national/release-detail?ReleaseID =3666.

156 On Election Day: "National Exit Polls: How Different Groups Voted," 2020 presidential exit polls conducted by Edison Research for the National Election Pool, *New York Times*, https://www.nytimes.com /interactive/2020/11/03/us/elections/exit-polls-president.html.

158 It has been four hundred years: "African Americans at Jamestown," National Park Service, https://www.nps.gov/jame /learn/historyculture/african-americans-at-jamestown.htm.

159 "No. I will never say": Footage of Malcolm X talking to a reporter, https://www.youtube.com/watch?v=cReCQE8B5nY.

159 "You've always told": *James Baldwin: The Price of the Ticket*, directed by Karen Thorsen (1989), https://www.youtube.com/watch?v=OCUlE5ldPvM.

160 "You got the first": Xuan Thai and Ted Barrett, "Biden's Description of Obama Draws Scrutiny," CNN, February 9, 2007, https://www.cnn.com/2007/POLITICS/01/31/biden.obama/.

162 Coleman was located: 1910 Louisiana Census, U.S. Census Bureau, https://www2.census.gov/library/publications/decennial/1910/abstract /supplement-la.pdf.

162 The town was situated: Gaytha Thompson, "History of Bienville Parish," in *Biographical and Historical Memoirs of Northwest Louisiana* (Chicago and Nashville: The Southern Publishing Company, 1890), http://files.usgwarchives.net/la/bienville/history/hist2.txt.

162 "Poor God-fearing people": https://africanamericanhigh schoolsinlouisianabefore1970.files.wordpress.com/2017/10/coleman -college-definitive.pdf.

163 Washington stood: "Coleman College," America's Lost Colleges, https://www.lostcolleges.com/coleman-college.

163 "There has never been": William Anthony Aery, "Loosening Up Louisiana," *The Survey* 34, no. 12 (June 15, 1915): 267.

164 "like history written": Mark E. Benbow, "Birth of a Quotation: Woodrow Wilson and 'Like Writing History with Lightning,'" *Journal of the Gilded Age and Progressive Era* 9, no. 4 (October 2010): 509–33, https://doi.org/10.1017/S1537781400004242.

164 "was the first Democratic": Kenneth O'Reilly, *Nixon's Piano: Presidents and Racial Politics from Washington to Clinton* (New York: Free Press, 1995); also O'Reilly, "The Jim Crow Policies of Woodrow

NOTES

Wilson," *Journal of Blacks in Higher Education*, no. 17 (Autumn 1997): 117–21, https://doi.org/10.2307/2963252.

165 "Paternal ancestor of William": Lynelle Cowan Stevenson and Martha Stevenson Owen, "History of Bienville Parish, Vol. I," USGenWeb Archives, http://files.usgwarchives.net/la/bienville/history/book/vol1b.txt.

166 Southampton is the same: Terence McArdle, "The Day 30,000 White Supremacists in KKK Robes Marched in the Nation's Capital," *Washington Post*, August 11, 2018, https://www.washingtonpost.com/news/retropolis/wp/2017/08/17/the-day-30000-white-supremacists-in-kkk-robes-marched-in-the-nations-capital/?utm_term=.962d120fc151.

166 "Color distinctions mattered": Michael W. Fitzgerald, *Reconstruction in Alabama: From Civil War to Redemption in the Cotton South* (Baton Rouge: Louisiana State University Press, 2017), 109.

167 My ancestors: Elmore County Heritage Book Committee, *The Heritage of Elmore County, Alabama* (Clanton, AL: Heritage Pub. Consultants, 2002), 239.

167 It had been homesteaded: "Arkansas Properties on the National Register of Historic Places: Kiblah School, Doddridge, Miller County," *Arkansas Historic Preservation Program* (blog), January 20, 2019, https://www.arkansaspreservation.com/blog/arkansas-properties-on-the-national-register-of-historic-places-kiblah-school-doddridge-miller-count.

169 "By 1928 one in every": Mary S. Hoffschwelle, *Preserving Rosenwald Schools*, 2nd ed. (National Trust for Historic Preservation, 2012), https://forum.savingplaces.org/HigherLogic/System/Download DocumentFile.ashx?DocumentFileKey=693200ab-b3c9-7ee9-f177-6ed15bcd491b.

169 Washington was hesitantly: Booker T. Washington, "Atlanta Exposition Speech," September 18, 1895, address to the Cotton States and International Exposition in Atlanta, transcript, Library of Congress, https://memory.loc.gov/cgi-bin/ampage?collId=ody_mssmisc&fileName=ody/ody0605/ody0605page.db&recNum=0&it?loclr=blogtea.

170 "The wisest among my race": Ibid.

171 "I have long since": Booker T. Washington, *Up from Slavery: An Autobiography* (New York: Doubleday, Page & Company, 1901), 16.

171 "prevents the African-American": "Transcript: Barack Obama's

NOTES

Speech on Race," National Public Radio, March 18, 2008, https://www.npr.org/templates/story/story.php?storyId=88478467.

172 "went through the school": Washington, *Up from Slavery*, 16–17.

173 "Chum, I am a": Edith Armstrong Talbot, *Samuel Chapman Armstrong: A Biographical Study* (New York: Doubleday, Page & Company, 1904), 86.

173 "a great man": Washington, *Up from Slavery*, 54.

174 "The Tuskegee school": Ibid., 137.

174 "Populism": Joseph H. Taylor, "Populism and Disfranchisement in Alabama," *The Journal of Negro History* 34, no. 4 (October 1949): 410–27, https://www.jstor.org/stable/2715608?read-now=1&seq=1#page_scan_tab_contents.

175 "It is a pleasure": Washington, *Up from Slavery*, 169.

175 "I am opposed": James K. Vardaman, "A Governor Bitterly Opposes Negro Education," TeachingAmericanHistory.org, February 4, 1904, http://teachingamericanhistory.org/library/document/a-governor-bitterly-opposes-negro-education/.

176 "In 1863 the Negro": Video of Martin Luther King's "The Other America" speech at Stanford University in 1967, https://kinginstitute.stanford.edu/news/50-years-ago-martin-luther-king-jr-speaks-stanford-university.

177 "But not only did": https://youtu.be/pLV5y4utPKI.

177 "created an institutional vacuum": Jim Downs, "Emancipation, Sickness, and Death in the American Civil War," *Lancet* 380, no. 9854 (November 10, 2012), https://www.thelancet.com/journals/lancet/article/PIIS0140-6736(12)61937-0/fulltext.

177 Downs estimates: Jennifer Schuessler, "Liberation as Death Sentence," *New York Times*, June 10, 2012, https://www.nytimes.com/2012/06/11/books/sick-from-freedom-by-jim-downs-about-freed-slaves.html.

178 "Let me heartily": Letter from W. E. B. Du Bois to Booker T. Washington, 1895, Columbia University, https://www.college.columbia.edu/core/content/letter-web-du-bois-booker-t-washington-september-24-1895.

180 I was part of: Holly Peterson, "What Jeffrey Epstein's Black Book Tells Us about Manhattan," *Financial Times*, August 23, 2019, https://www.ft.com/content/eb4ec1ae-c405-11e9-a8e9-296ca66511c9.

NOTES

182 storied Black congressman: Tom Wicker, "Mississippi Delegates Withdraw, Rejecting a Seating Compromise; Convention Then Approves Plan," *New York Times*, August 25, 1964, https://archive.nytimes.com/www.nytimes.com/library/politics/camp/640826convention-dem-ra.html.

183 "the days of the 'Ku Klux'": Washington, *Up from Slavery*.

183 lone black delegate: R. L. Nave, "Voting Rights: Was Chief Justice Roberts Wrong About Voting in Mississippi?," *Jackson Free Press*, July 10, 2013, https://www.jacksonfreepress.com/news/2013/jul/10/voting-rights-was-chief-justice-roberts-wrong-abou/.

183 "won the Civil War": James W. Loewen, *Up a Creek, With a Paddle: Tales of Canoeing and Life* (Oakland, CA: PM Press, 2020).

184 "Negro suffrage is": "Judge Calhoon's Views," *Clarion-Ledger* (Jackson, MS), March 6, 1890, https://www.newspapers.com/clip/23870464/the-clarion-ledger/.

184 "It is the manifest intention": *Journal of the Proceedings of the Constitutional Convention of the State of Mississippi* (1890), 275.

185 "123,000 of my fellow-men": *Literary Digest*, 1890. Accessed via GoogleBooks.

186 Other southern states: Robert C. Smith, "Montgomery, Isaiah T." in *Encyclopedia of African-American Politics* (New York: Facts on File, 2003), 228–29.

186 As Frederick Douglass said: David Mark Silver, "Montgomery, Isaiah Thornton," in *Africana: The Encyclopedia of the African and African American Experience*, 2nd ed., vol. 3., eds. Kwame Anthony Appiah and Henry Louis Gates Jr. (New York: Oxford University Press, 2005), 56.

6. The Reunion

193 As Samaria would tell: Connie Schultz, "It's the Last Video I Have of My Child Alive," *Politico*, October 15, 2015, https://www.politico.com/magazine/story/2015/10/tamir-rice-mother-interview-213252.

193 His urn: Charles M. Blow, "I Interviewed Samaria Rice Today," Facebook, July 20, 2016, https://www.facebook.com/CharlesMBlow/photos/a.91950624988/10154954014949989/?type=3&theater.

194 While the city of Cleveland: "Race and Ethnicity in Cudell, Cleveland, Ohio," Statistical Atlas, https://statisticalatlas.com/neighborhood/Ohio/Cleveland/Cudell/Race-and-Ethnicity.

NOTES

196 With such a large family: "2020 Health & Human Services Poverty Guidelines / Federal Poverty Levels," Paying for Senior Care, August 23, 2020, https://www.payingforseniorcare.com/federal-poverty-level.

203 There were five of us: Morgan State University School of Global Journalism and Communication staff directory, https://www.morgan.edu/school_of_global_journalism_and_communication/our_staff/dewayne_wickham.html.

203 April D. Ryan: Richard Prince, "5 Black Journalists Were Allowed to Ask President Obama Questions on Air Force One before His Selma Speech," Black Christian News, March 10, 2015, https://blackchristiannews.com/2015/03/5-black-journalists-allowed-ask-president-obama-questions-air-force-one-selma-speech/.

204 "For black people": Jemar Tisby, "I'm a Black Man Who Moved to the Deep South. Here's What It's Teaching Me about Race," *Vox*, January 4, 2019, https://www.vox.com/first-person/2017/10/31/16571238/black-man-deep-south-race.

206 There are the buildings: Inge Oosterhoff, "Under Wall Street Lies a Legacy of Slavery. High Time for a Tour," Correspondent, October 1, 2016, https://thecorrespondent.com/5312/under-wall-street-lies-a-legacy-of-slavery-high-time-for-a-tour/715844997824-125573a2.

206 settlement called "Little Africa": Andrew Berman, "Black History in Greenwich Village: 15 Sites Related to Pioneering African-Americans," 6sqft.com, February 16, 2018, https://www.6sqft.com/Black-history-in-greenwich-village-15-sites-related-to-pioneering-african-americans/.

206 There was the Black: Sam Neubauer, "San Juan Hill: The Upper West Side's Lost Neighborhood," ILoveTheUpperWestSide.com, July 18, 2020, https://ilovetheupperwestside.com/san-juan-hill-lincoln-center/.

207 "produce 70 percent": Alexander Nazaryan, "Black Colleges Matter," *Newsweek*, August 18, 2015, https://www.newsweek.com/black-colleges-matter-363667.

208 "I find, in being black": *Purlie Victorious*, film, 1963, https://www.youtube.com/watch?v=wj6dH82LXQE.